D0057088

ICE, MUD AND BLOOD

Lessons from climates past

Chris Turney

Macmillan

London   New York   Melbourne   Hong Kong

© Chris Turney 2008

All rights reserved. No reproduction, copy or transmission of this publication may be made without written permission.

No paragraph of this publication may be reproduced, copied or transmitted save with written permission or in accordance with the provisions of the Copyright, Designs and Patents Act 1988, or under the terms of any licence permitting limited copying issued by the Copyright Licensing Agency, 90 Tottenham Court Road, London W1T 4LP.

Any person who does any unauthorised act in relation to this publication may be liable to criminal prosecution and civil claims for damages.

The author has asserted his right to be identified as the author of this work in accordance with the Copyright, Designs and Patents Act 1988.

First published 2008 by
Macmillan
Houndmills, Basingstoke, Hampshire RG21 6XS and
175 Fifth Avenue, New York, N. Y. 10010
Companies and representatives throughout the world

ISBN-13: 978–0–230–55382–8
ISBN-10: 0–230–55382–6

This book is printed on paper suitable for recycling and made from fully managed and sustained forest sources. Logging, pulping and manufacturing processes are expected to conform to the environmental regulatons of the country of origin.

A catalogue record for this book is available from the British Library.

A catalog record for this book is available from the Library of Congress.

10   9   8   7   6   5   4   3   2   1
17   16   15   14   13   12   11   10   09   08

Printed and bound in China

QL
861.3
T87
. 2008

To Cara and Robert: may you inherit a better world

CONTENTS

LIST OF FIGURES

LIST OF PERMISSIONS AND FIGURE SOURCES

Figure 2.2 'The continent of Rodinia 750 million years ago' was adapted from Torsvik, T. H. (2003) The Rodinia jigsaw puzzle. *Science*, **300**, 1379–81.

Figure 4.1 'Ice flow in the Greenland Ice Sheet' was adapted from Dansgaard, W. *et al.* (1969) One thousand centuries of climatic record from Camp Century on the Greenland ice sheet. *Science*, **166**, 377–81.

Figure 5.3 'Thermohaline circulation' was adapted from Lowe, J. J. and Walker, M. J. C. (1997) *Reconstructing Quaternary Environments* (Longman, Harlow) and Henson, R. (2006) *The Rough Guide to Climate Change* (Rough Guides, London).

Figure 7.1 'Rebounding land after the ice' was adapted from Morén, L. *et al.* (2001) Climate and shoreline in Sweden during Weichsel and the next 150,000 years. *Technical Report TR-01-19*, Swedish Nuclear Fuel and Waste Management Co.

Figure 7.2, 'The spread of European farming', was adapted from Turney, C. S. M. and Brown, H. (2007) Catastrophic early Holocene sea level rise, human migration and the Neolithic transition in Europe. *Quaternary Science Reviews*, **26**, 2036–41.

Map backgrounds from Maps in Minutes, World Countries 2005, ESRI Data and Maps Media Kit.

ACKNOWLEDGEMENTS

A book on this vast and urgent topic wouldn't have been possible without the help and guidance of a lot of friends and colleagues. During my career I've been extremely fortunate to work with some fantastic researchers, teachers and mentors who have taught me the value of the past. A huge number of people have helped me gain a small understanding of how the world works. As a younger researcher I soon learnt that many of the best discussions were to be had in the pub. I still owe several drinks to Mike Baillie, Tim Barrows, Michael Bird, Nick Branch, Chris Caseldine, John Chappell, Ed Cook, Russell Coope, Keith Fifield, Doug Harkness, Konrad Hughen, Sigfùs Johnsen, Niek de Jonge, Peter Kershaw, John Lowe, Matt McGlone, Colin Murray-Wallace, Jonathan Palmer, Sune Rasmussen, Bert Roberts, Jim Rose, Ian Shennan, Dave Stainforth, Jørgen Peder Steffensen, Chris Stringer, Jim Teller, Mike Walker, and Nikki and Alan Williams. If I've forgotten anyone else, please forgive me. If I've got anything wrong, it is most certainly my fault.

I'd like to thank Sara Abdulla who helped get this book off the ground and Alex Dawe at Macmillan for her support while writing. Heidi Brown kindly helped draw Figure 7.2.

My family have been hugely important to me in the preparation and completion of this book. My parents Ian and Cathy Turney have always inspired me to enjoy and understand the environment, for which I am eternally grateful.

I've written this book wherever I could; mostly in a small room at the back of the house in the small hours. It took far longer than I ever imagined possible and would never have been finished without a huge amount of patience, support and ideas from my gorgeous wife Annette. Thank you darling. I promise this is the last time...

Finally, I'd like to dedicate this book to my two children Cara and Robert. I hope my generation make the changes needed. You deserve a world in better shape.

INTRODUCTION

Imagine a world of wildly escalating temperatures, apocalyptic flooding, devastating storms and catastrophic sea level rise. This might sound like a prediction for the future or the storyline of a new Hollywood blockbuster but it's something quite different: it's our past. In a day and age when we're bombarded with worrying forecasts for future climate, it seems hard to believe that such things could come to pass. Yet almost everywhere we turn, the landscape is screaming out that the world is a capricious place. The problem is that if we don't tune in, the message is lost. We need to decipher the past and learn from it.

I came to realize the value of the past relatively late. In early 1994 I was at a crossroads. I was finishing a degree in Environmental Sciences and knew I wanted to stay in science, but what I should specialize in daunted me. There was so much exciting work being done. In spite of all the choice, the world's changing climate was particularly appealing. It was a thrilling time to get involved. Only six years before, some remarkable events had taken climate change from being a largely fringe issue to headline news. In 1988, James Hansen, the director of NASA's Goddard Institute for Space Studies, had testified before the US Congress that global warming was upon us. Across the Atlantic, the British Prime Minister Margaret Thatcher had famously warned the Royal Society we had a problem. Climate was suddenly an international issue and there was a real sense of urgency: science had to find out more. I could make a contribution by giving a long-term perspective. I could help find out what had gone before.

Before I knew it I was a PhD student knee deep in a water-filled trench in southwest Ireland. I was with two fellow students from the University of London, digging in a peat bog, trying to reach the ancient lake sediments that lay at the bottom. 12,000 years earlier, this green and very wet land had been much more like the Arctic of today. Although no great ice sheet would have lain nearby, it would have been very much colder. The question was by how much? After four days of continuous rain, I was fast losing interest. We had real trouble getting the water out of the pitifully shallow trench we had dug. No matter how hard our pump worked, the pit just kept filling. My entire body was soaked through and I was numb with cold. Yet as we laboured each day to salvage something from the exercise, my eyes were drawn to the walls of peat that surrounded me. Poking out from the sides were plant remains: small branches, leaves and seeds; an ecosystem preserved over thousands of years. In the fleeting moments when we did seem to be winning the battle with the water table I could see my feet; floating out of the clay at the bottom of the trench were fragments of glittering insect remains. I picked one up. Despite the worsening conditions, an enthusiastic voice shouted over the din of the pouring rain: 'It's a beetle'. It was *Diacheila arctica*, a species of beetle that likes summer temperatures 4 °C cooler than where we stood. The remains we had uncovered had lived twelve millennia ago, yet its descendants are no longer found in Ireland. It was clear evidence that the climate had changed. I had got the bug. Past climates it was.

The past has a fabulously important role to play in helping us understand what our planet is capable of. Only from bygone ages can we get a tangible handle on what has gone before and what might happen again. As we are increasingly alerted to the perils of climate change, we need a long-term perspective to evaluate the risks and make an informed judgement on how best to manage them. We could try asking family, friends, or any other poor unsuspecting souls what they remember of their past: has the climate changed since they were kids? The chances are they'll reply yes. Unfortunately, the memories we collect as individuals over a lifetime are pretty suspect. As children, we spend much of our time outside, exposed to the highs and lows of the elements. As we get older, most of us pass ever more of our lives indoors, isolated from the extremes by our atmosphere-controlled cars, homes and workplaces. We become increasingly detached from the world outside. We're hardly reliable sources for chronicling climate change. Where our family memories become less reliable, records from the past provide a reality check, but the problem with all this is that regardless of where we live on this planet, the longest continuous weather records stretch back only a few centuries.

Ideally, if we wanted to be fully alert to a changing climate, we'd have weather stations stacked up around the world, recording different aspects of the elements over millennia. Sadly we're not even close to this. Today's coverage of weather stations is somewhat erratic. Some parts of the globe are well served: the United Kingdom, for instance, has somewhere of the order of 5,000 official rainfall stations, with several hundred of them also collecting temperature. It also has the longest continuous dataset in the world; the Central England temperature record has monthly averages stretching back to 1659, with daily changes logged since 1772. In contrast, Africa, which makes up around a fifth of the world's land surface, has only around 1150 stations; about a fifth of what is needed to do a proper job. The longest continuous record in Africa is at the Royal Observatory in Cape Town which goes back no further than 1842. As a result, we're not ideally placed to get a full picture of the world's climate over time.

In some parts of the world we're helped by a few early snapshots of the weather recorded by enthusiasts. In the 18th century, an interest in meteorology was highly applauded. You didn't have to be hairy and middle aged. A clean-shaven and young Thomas Jefferson kept a weather journal; from this we know that it was pretty mild in Philadelphia during the Declaration of Independence on 4 July in 1776. We can go further back by looking at pictures where the weather forms part of the scene. These are notoriously difficult to interpret and should often be taken with a pinch of salt. For example, there's a beautiful painting by Emanuel Gottlieb Leutze showing George Washington crossing the Delaware River in 1776 to attack the British-backed Hessian troops stationed in Trenton, New Jersey. The river is clogged with enormous berg-like blocks of ice. It's a romantic piece of work that was painted some 75 years after the event but suggests a winter that would be unimaginable today. Sadly though, paintings can only take you back so far.

Happily, when our ancestors were fighting one another, nature was recording the conditions at the time; depending on the season, trees, peat bogs, ice and mud all preserve an archive of what the conditions were like when they were formed. The trick is how to read what nature has left behind.

We now know that all sorts of extreme and rapid changes have taken place in the dim and not so distant past; these have had major effects – both good and bad – on cultures and civilizations. As science gets a better feeling for how our planet works, it's becoming ever more clear that nowhere is really isolated from anywhere else. From the Alps to the Andes, seemingly unrelated parts of the world are connected in one way or another. What I hope is that, by reading on, you'll realize that we're starting to see a time in which we'll face challenges beyond anything our species has had to contend with before.

In *Ice, Mud and Blood* we're going to look at climates of the past. But to put these changes in some sort of context it's important to understand what today's fuss is all about. Why is there so much concern about climate change? To cut a moderately long story short, certain gases – known as greenhouse gases – trap the sunlight that's been reflected off the Earth's surface. Today these gases make up around 1 per cent of the atmosphere. This doesn't sound a lot, but without them our planet would be a rather chilly −18 °C instead of the balmy 14 °C we enjoy today. The problem isn't that these gases exist; the nub of the problem is there is an ever-increasing amount of them in the atmosphere. If people weren't screwing up the environment, the natural rate at which these gases are created would be around the same as the rate they're destroyed or locked up. As we'll see later, where this balance has been disrupted in the past it's caused all sorts of problems and these give us important insights into what the future might bring. The bottom line is that we're now running an unintended experiment that's threatening to go belly-up. Because of our seemingly insatiable demand for fossil fuels, we're pumping vast amounts of greenhouse gases into the atmosphere using carbon that was locked away millions of years ago. More reflected heat is getting trapped and the atmosphere is warming.

Predicting the effects of increasing amounts of greenhouse gases on the climate is not child's play, however. People are quite wary of forecasts for the future, and for good reason. The word 'forecast' itself often raises associations with soothsayers, astrologers and occasionally sacrificing the odd virgin. Indeed, some of the earliest predictions of what the weather might bring were based on the supposed influence of the stars, planets and Moon on the Earth's atmosphere. Now known as astrometeorology, this quack science reached a pinnacle of popularity in the 19th century. Probably one of its best known proponents was Patrick Murphy M.N.S. – Member of No Society – who was fascinated by how planetary bodies could influence our atmosphere via gravity, electricity and magnetism. He was a regular correspondent to *The Times* in the 1830s, warning of hazards and confirming any apparent successes in subsequent letters. In 1838, he published *The Weather Almanack (on Scientific Principles Showing the State of the Weather for every Day of the Year 1838)* in which he predicted that 20 January would be the coldest day of the year. When by happy chance it was, the book became a runaway success, reaching the heady heights of 45 editions in one year and spawning companion volumes. But not everyone was convinced. *The Times* felt moved to comment that Murphy had made '*some very happy guesses*'; poems and cartoons were penned in mockery; and a one-act farce was put on at the Sadlers Wells Theatre in the same year. Despite his best efforts, Murphy could not replicate his early success and the idea gradually fell by the wayside.

In spite of all the problems, the public were hungry to know what the weather might bring. It is here that one of science's most enigmatic characters comes into our story: Robert Fitzroy. Probably best known as captain of *HMS Beagle* when it took a relatively unknown Charles Darwin round the world on a voyage of scientific discovery between 1831 and 1836, Fitzroy later managed to squeeze in some time as a Member of Parliament and be the second governor of New Zealand. A devoutly religious man, he struggled through much of his life with the knowledge that he had given Darwin the inspiration to write *Origin of Species* and had inadvertently helped let the evolutionary cat out of the bag. In 1854, the British Meteorological Office was set up to collect weather data and Fitzroy was appointed its head. By 1860, he had convinced the authorities that a network of observation stations around the British Isles could give an early warning of impending storms. It was not a big leap of faith to go a step further and provide the first national weather forecasts. These began a year later and were published in *The Times*; Fitzroy himself actually first coined the term 'weather forecast'. Although this was pioneering stuff, it wasn't long before readers and the newspaper were criticizing failed forecasts. Fitzroy was almost too honest about the problems, and a year after taking his own life – apparently in a fit of depression – the forecasts were stopped; they didn't come back until another decade had passed.

At the turn of the 20th century, a British scientist called Lewis Fry Richardson first tried to predict the changing weather using known physical principles. The results were not a glowing success – in fact they were dreadful – but laid the groundwork for the models used today to predict the weather. Computer models can now produce weather forecasts up to five days into the future, using known conditions at the time and then predicting the movement of air and its many characteristics (including heat and moisture). Unfortunately, there's a limit to how far ahead you can forecast. You might have heard of the idea of a butterfly flapping its wings in one part of the world and causing a hurricane or tornado in another. It's not to be taken literally, but the principle is that small changes can feed through any dynamic system and cause large changes some time later. The atmosphere is essentially chaotic. The result is that small changes at the beginning of a five-day forecast can feed through and disproportionately change what the weather does. One day a strong wind might be heading out to sea, the next day it's set its sights on a densely populated city. There's not a lot that can be done about this apart from constantly updating and revising the predictions; the world is just too complicated for forecasts to hold true for weeks on end.

But all is not lost for climate prediction. There's a major difference between weather and climate models. A model producing a weather forecast will give a prediction for what the conditions will be like in different parts of the world just a few days into the future. With climate, we're not interested in the short-term changes in meteorological conditions. It's the long-term we're after. The short-term chaos we see with weather forecasts tends to smooth out over decades and centuries. As a result, we can get a handle on what will happen on average in different parts of the world in the future. A climate model will not give the daily temperature and rainfall for each day of the year in different parts of the world over a hundred years. The key thing is that it will give an idea of the average conditions. It's the same principle as the changing seasons and their effect on temperature; if we live in the northern hemisphere, we don't know what the temperature will be for every day in July next year, but we know that it will be warmer on average than December.

A good way to picture the above is to consider the changing popularity of some children's names. From one year to the next, you can guarantee that a number of parents will name their hapless offspring after a film star or sport hero. Most of these poor individuals then carry this burden all through their lives; a symbol of an earlier age. It's hard to say for sure what the names will be from one year to the next but you know they'll change eventually. Foul deeds are done at the font. But just as it's a sure thing that different dreadful names will be bequeathed by well-meaning parents over time, we also know that climate can change. If a celebrity remains a household name, there will be some sort of stability. But as soon as a new megastar comes on the scene, the whole system shifts and another generation of children are named something wacky. With climate, it's the change in forcing that drives long-term weather.

Over the past century, weather stations have shown the world's average temperature has increased by around 0.8 °C. But it's not been one long steady rise. An increase in temperature up to the 1940s was interrupted by cooling during the 1950s through to the 1970s, followed by a steep rise until today. Keep in mind that this is an average; not everywhere was warming by the same amount. Importantly, we can use climate models to get a handle on what combination of factors drove the temperature trend. It now appears that most of the warming up until the 1940s can be explained by changes in the heat coming from the Sun, while the cooling seems to have been principally caused by sunlight-reflecting liquid particles of sulphate produced by volcanic eruptions and the burning of fossil fuels. It's only by adding greenhouse gases to the models that we can explain the dramatic warming seen since the 1970s; the levels in the atmosphere have now become so high that they appear to have overridden what should be a cooling trend. Ultimately, this is why most scientists are convinced that people have pushed the climate out of its natural state.

Seeing this alarming trend in the late 1980s, the World Meteorological Organization and the United Nations set up the Intergovernmental Panel on Climate Change – better known as the IPCC – to look into the scientific basis of human-induced climate change. The IPCC is now a veteran. In 2007 it presented its Fourth Report, written by over 800 contributing authors, reviewed by more than 2,500 scientific expert reviewers and representing over 130 countries. The result is very much a consensus and it took six years to get it. It contains a worrying vision of a climate out of control. We'll look at what the IPCC said in more detail later, but suffice to say that, depending on what the world's carbon dioxide levels do in the future, predictions for 2100 suggest temperatures are likely to be somewhere between 1.1 °C and 6.4 °C warmer than 1990, with sea level 20 to 60 cm higher. Worryingly, the upper limits of these predictions don't include the full industrialization of China and India, implying that things could get even worse.

But is the warming over the past few decades unprecedented and beyond what's natural? Climate models are giving us a good idea of what is driving the change and an indication of what might happen in the future. But are these projections realistic? Should we get out of any coastal real estate pronto and run to the hills? Fortunately the past has seen many of the changes we're likely to see in the future. Some of the changes have been catastrophic. If we carry on behaving as we are, the future doesn't look rosy.

When I originally started to research this book, I wanted to show how we can piece together the past. I soon discovered, however, that when I told a friend or family member some new fact I'd unearthed I was often met with shock and surprise. It was news to them. Yet it's been bread and butter to scientists for some years now. Not enough of our work gets out to the public.

Misunderstandings about science really hit me between the eyes in July 2007 when I attended the Australian screening of a rather strange 'documentary' called *The Great Global Warming Swindle*. This film contains a bizarre mixture of half-truths, misinformation and fabrications to argue that changes in the climate seen today are not caused by human activity but are solely the result of the Sun. To make some sort of case, the program tied itself up in all sorts of knots, often falling back on results that were over 15 years old. Essentially the broadcast tried to argue there's nothing to worry about, despite the plethora of evidence that shows otherwise.

There was a lot of controversy when the Australian Broadcasting Corporation purchased the wretched thing, but to their credit they set up an interview with the director, had a panel of experts and business people discuss the content and then opened up to questions from a live audience. As I sat through the film listening to the nonsense that was being spouted, I suddenly became aware that rather than being angered, many in the audience were sympathetic to the arguments. Once the questions began, I suddenly realized that many of my companions were either loonies or had been very badly informed. It struck home just how poor a job we've done as scientists in communicating our work. For an issue that is so important and needs people to make some very serious and far-reaching decisions, I saw that we still had a lot of work to do. Somehow, amongst all the other news jostling for attention, the science has almost got lost in the maelstrom.

In truth, there's probably fault on both sides. Scientists are a mixed bag when it comes to describing their work and its relevance. Add to this some political and industrial leaders obsessed with economic growth and you can end up with a very confused message. There's a very real danger that climate change is becoming just another scare story. Unfortunately, sticking our head in the sand won't make the problem go away. Despair isn't much use either. We can still sort the problem but we need to act now, individually and globally. Nations need to take a lead, but we can all play a role.

In this book I'm not going to describe ways in which we might reduce the amount of greenhouse gases we're emitting. There are already some excellent books available on this topic: George Monbiot's *Heat* and Dave Reay's *Climate Change Begins at Home* are great examples. What I want to show is how we know what happened in the past, how these changes came about, what is natural and what this all means for our future. As you'll see, if we're going to understand the future, we have to get a handle on what has gone before. We've been living on borrowed time, but we still have a choice. We can do something about the mess we've created. We can listen to the warnings from the past and change our ways.

Chapter 1
GREENHOUSE

They'd lock you up and throw away the key if you wanted to pump billions of tonnes of carbon into the atmosphere to see how it affected climate. Yet, seemingly without a care in the world, we're doing just that. Our voracious appetite for coal, gas and oil is causing somewhere on the order of 32 billion tonnes of carbon dioxide to be dumped into the air each year. By digging up fossil fuels and releasing their trapped energy, we're putting carbon back into the atmosphere that hasn't seen the light of day for millions of years. Carbon has freed us from living hand to mouth. Since our species evolved over 100,000 years ago, our numbers have ballooned to six and a half billion. To a large extent it's all thanks to carbon. Carbon has become our liberator, fuelling our economy and lifestyle.

There is no doubt that a little bit of carbon in the atmosphere is good for us; it helps give our planet the pleasantly warm temperature it has today. The problem is we're tipping the balance. We're on a fast track to an atmosphere that hasn't been witnessed for some 55 million years. Although something of this age might sound so hopelessly ancient as to be of no relevance to today, there's actually a very important lesson here: the fallout from this last flirtation with a carbon-rich atmosphere took thousands of years to sort out.

To find out what happened at this time we can look in the sea, or more precisely, under it. Some of the largest research ships in the world have only one purpose: to take very long, thin cores through the ocean bed. Some of these cores capture fine muds that were laid down on the sea floor over tens of millions of years. After the novelty of looking at something different has worn off, the chances are most of the cores wouldn't impress you very much; one fine mud looks pretty much like any other fine mud. But it's what's in the mud that counts. When we delve into layers from 55 million years ago, something disturbing emerges. In what appears to be a blink of an eye, the temperature of this planet went through the roof. This is the mother of all doomsday scenarios. This was the Palaeocene–Eocene Thermal Maximum (often shortened to PETM).

Back in 1991, James Kennett at the University of California and Lowell Stott at the University of Southern California reported results from a core taken in the Weddell Sea off Antarctica that spanned hundreds of thousands of years. Within the core they found the remains of minuscule ocean-dwelling organisms known as foraminifera – affectionately called 'forams' – that lived at the surface and bottom of the oceans. When alive, forams take up calcium, carbon and oxygen from the seawater to form calcium carbonate shells that come in all sorts of weird and wacky shapes and sizes: some are enormous – by foram standards – and can measure up to 20 centimetres; others can be as small as 0.1 millimetres across. They're also remarkably diverse; there are somewhere around 4,000 species living within different ecological niches in the sea. But these small organisms don't have long for this world; they typically shuffle off their mortal coil within a few weeks. The important thing is that once the forams die, the shells fall to the seabed and get buried by ocean muds. Because different forams are found through the world's oceans, they can give a great idea of what the conditions were like at a particular moment in time.

Looking within the foram carbonate shells, Kennett and Stott measured the amount of two forms of carbon known as carbon-12 and carbon-13. These different numbered versions of carbon are known as isotopes, and although they have the same chemical properties as one another, the carbon-13 is slightly heavier than its companion. Although it might not sound much, subtle changes in the ratio of these two versions of carbon can give important insights into what happened in the past.

What Kennett and Stott noticed was that across the geological boundary separating the Palaeocene and Eocene there was a huge shift in the carbon isotopic makeup of the shells: there was suddenly a lot more carbon-12, a shift sometimes described as going 'negative'. This was unusual. The sort of change they found was not what you'd normally expect to find in forams. Not only that, but the relationship between those living on the surface – the so-called planktonic forams – and those living at the bottom – the benthics – went askew around 55 million years ago. Under normal conditions, the surface dwellers tend to have more carbon-13 than their comrades in the depths. Given the choice, living organisms prefer using the lighter carbon for growth because it requires less energy to build organic matter. At the surface there are often intense blooms as forams and other living organisms go in for a frenzy of growth. The result is that most of the carbon-12 is used up, leaving only the heavier isotope for making the carbonate shell. But in the depths, the water gets a regular flurry of dead organic matter from the surface (Figure 1.1). There's a lot less life down there, so the forams can be more choosy in building their shells with the lighter carbon. Across the Palaeocene–Eocene boundary, this traditional difference disappeared, and regardless of where in the ocean the forams lived, the shells all contained more of the lighter carbon.

Other things also suggest that this time was topsy-turvy. Alongside the big shift in carbon isotopes was a mass extinction of benthic foraminifera. Somewhere between 35 and 50% of all benthic forams were lost at this time, making it one of the largest extinctions of the past 90 million years. Something out of the ordinary must have happened. Kennett and Stott pointed the finger at massive global warming. As the seawater temperatures rose, they argued, the amount of oxygen reaching the bottom of the world's oceans crashed. It was a recipe for extinction on a colossal scale.

We can get a handle on what the temperatures were actually doing by looking at a particular type of organic compound

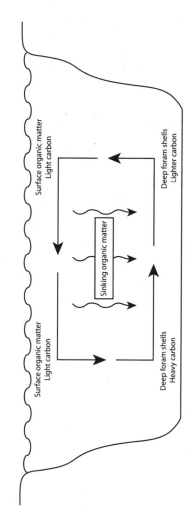

Figure 1.1 Changing carbon isotopes in forams.

produced by a micro-organism called crenarchaeota. These uninteresting looking organisms are a type of bacteria that are found in oceans and lakes. They might not be very easy on the eye, but crenarcheota are a gold mine when it comes to reconstructing past climates. During growth, the water temperature controls the production of a lipid called tetraether in their membrane. When the little critters die this lipid falls to the sea floor and becomes locked up in the sediment. By measuring the tetraether within sediments that were laid down in the past, it's possible to come up with a temperature scale called TEX$_{86}$. To simplify matters, I'll just call it the TEX scale. The key thing is that the TEX result from 55 million-year-old Arctic sediments shows that temperatures rose to as high as 25 °C. But what does this number mean?

Climate implies average conditions. But it's often not possible to reconstruct past changes at the daily, weekly or monthly level. In most cases, a proxy of the conditions will give us at best a decadal average, and often for just one part of the year, such as the growing season. Unfortunately, averages are far from the whole story. They tend to paint an image with a very broad brush. If the weather is largely unchanging, then the average can be thought of as the 'climate'. But this is often not the case. Just imagine two different years for where you live. In one year not a lot happens. It rains through the year, farmers produce a reasonable amount of food and generally speaking most people are happy. Fast forward a few years, and consider a year where for most of the time no rain falls. Drought ensues, but during one day, a whole year's worth of rain falls. Taken as an average, the amount of water that fell on the land was the same, but the pattern couldn't have been more different. It's pretty hard to pick up this sort of difference in the past, particularly when you're dealing with time-scales of millions of years. The best we can hope to do is sample at the highest resolution possible. But even then we're dependent on what sort of record we're looking at. We'll see later how some proxies – such as trees, corals and ice – can be used to find out what happened year by year, but in the meantime just bear in mind that some reconstructions might be skewed to the average of one season over several years.

This is not the disaster it might at first appear; it's important to try understand how the seasons behaved in the past. In the case of the TEX reconstruction through the Paleocene–Eocene, most of the lipids being put down on the sea bed would have been made when there were blooms on the surface, suggesting that the temperature of 25 °C probably represents summer rather than an annual average. Even so, it must have been pretty warm all year round because the poles were free of ice. The tropics, which tend to be less prone to big temperature changes, warmed by 5 °C. The evidence all seems to point to the world experiencing a devastating hothouse. It's this period of warmth that's been dubbed the Palaeocene–Eocene Thermal Maximum.

In different parts of the world there were massive changes in the flora and fauna. If you had been able to visit the Norwegian archipelago of Svalbard at the time, you'd have found a climate similar to today's Florida, with crocodile-like beasts and plant-eating creatures living in a temperate rainforest. Turtles thrived on Ellesmere Island in the Canadian Arctic. In the east Russian peninsula of Kamchatka, palm trees grew. Whereas before, most mammal groups had been confined to Asia, during and after the warming many of these migrated into Europe and North America. Clearly big changes were put in train by the warming.

The PETM is now thought to have occurred over 160,000 years or so, with the big shift to lighter carbon taking place over some 30,000 years. It might well be that the changes were faster than this, but with ocean sediments spanning such a distant period of time it's hard to tell. As a result, I'll stick with the best, albeit conservative, estimate we have.

The warming in the deep gave a massive hint that something had happened to the ocean's circulation. Today's ocean takes warm surface waters from the tropics up to high latitudes; evaporation concentrates the sea salt and the water progressively cools. Eventually the water becomes dense enough to sink, forming deepwater which then heads equatorwards through the ocean depths. The winds and tides mix up the ocean, allowing the deepwater to well back up to the surface, where it is then warmed enough to flow polewards and repeat the whole process. We'll look at this more closely later on, but for our purposes this description will suffice. During the Palaeocene it looks like something similar to today's circulation was in operation. The major uncertainty was whether it changed once the warming kicked in.

Flavia Nunes and Richard Norris of the University of California hit upon a rather neat solution to work out what happened in the ocean. Typically, the longer any deepwater is separated from the surface, the more nutrients it collects, largely because dead and decaying material falls to the ocean floor. Those benthic forams living under the sites of deepwater formation are bathed in seawater that has only just left the surface. As a result they tend to have little of the light carbon because most of it has been used up at the upper levels. But the further you go along the flow of deepwater, the forams progressively gain more of the lighter carbon from the surface. Taking 14 deep-sea sites from around the world, Nunes and Norris measured the changing ratio of the different isotopes of carbon in the shells of benthic forams. What they found was stunning. Before and after the PETM, the sites with the most carbon-13 were in the high latitudes of the southern hemisphere, suggesting that most of the ocean deepwater was being formed there. But within the first 5,000 years of the start of the PETM, there was a major change: the northern hemisphere benthic forams suddenly showed more carbon-13 than their southern counterparts. For at least 40,000 years, it looks like deepwater suddenly switched from being made in the south to the north. It then took another 100,000 years before circulation reverted to its previous state and business as usual was re-established.

But the key to the conundrum of what had caused the big upset was published just a year after Kennett and Lowell's first results. In 1992, Paul Koch and colleagues at the Carnegie Institution of

Washington showed that the carbon isotope shift within the ocean also took place on land. Animal teeth and carbonates found in soils dating to 55 million years ago showed a similar big shift to lighter carbon. Whatever had caused the big change in carbon was not unique to the ocean. It had to have affected the whole carbon cycle. The most obvious culprit was a huge pulse of gas. But how could a gas be the cause of such a big change?

Virtually all the energy that drives the world's climate comes to us from the Sun. Each day we're bathed in the stuff. It drives the oceans and the atmosphere; it makes life possible. Without it our planet would be a desolate hole. But this energy doesn't come to us in one form; it's a package of different sized waves. Disconcertingly, these waves are known as radiation, but they have nothing to do with radioactive decay. The term radiation simply comes from the Latin word *radiare*, which means to emit rays.

To understand climate and the changes that are happening we need to look briefly at how the Sun's energy gets to us. A good way to visualize this is to throw a stone into your local duck pond: assuming you miss the local wildlife you'll see a series of ripples move out from the centre, each made up of a peak and a trough. Now if you throw lots of different sized stones onto the same point, lots of different sized waves will be created. As the waves travel across the pond's surface they transfer energy from the centre to the edge. In much the same way, the energy from the Sun comes to us with a range of different sized waves; it's just that we can't see all of them.

We're most familiar with the wavelength of maximum radiation from the Sun: visible light. Because it dominates, our bodies have evolved to sense its energy. The size of the waves of sunlight, however, is a lot smaller than our efforts in the pond; the distance from peak to peak (or trough to trough) is only 0.0004 to 0.0007 millimetres. Other wavelengths of energy produced by the Sun need specialized equipment to be detected. Of particular importance to our story are ultraviolet light – often referred to as UV – which has a slightly smaller wavelength than visible light, and infrared light which has a somewhat longer wavelength. Of the 64 million watts per square metre that the Sun produces, the same area at the top of our atmosphere sees only around 1366 watts (the so-called solar constant). Crucially, not all this energy reaches the surface. Some, for instance, is scattered by particles in the air. When you look up at the sky on a cloudless day you're inadvertently watching this process; of all the colours in the spectrum, blue is the most strongly scattered, dominating the sky. On average, however, about 30% gets reflected back out to space by dust and clouds in the air and off the ground. A further 20% is absorbed in the air by water vapour and clouds, while the remaining 50% is absorbed by the Earth's surface, causing warming.

It's important to realize that any object with a temperature above absolute zero (defined as −273 °C) gives off energy. As a result, virtually everything does so. Our bodies are warmer than absolute zero and therefore give off energy. This book is above absolute zero, so it too is giving off energy. The Sun's surface temperature is around 6000 °C and radiates a lot of energy, most of it at very short wavelengths. Our planet, though, is much cooler than the Sun, so most of the energy absorbed at the surface and by the atmosphere is stretched out and given back off as infrared. Not all of this infrared energy goes straight out into space, however; some is intercepted by clouds and, importantly for our story, gases in the atmosphere.

The idea that our planet's atmosphere may help keep things warm was first proposed by the French scientist Joseph Fourier; in 1827 he used the analogy of a 'glass bowl' and suggested that gases in the air let the sunlight through but kept the infrared in, heating up the Earth's surface temperature. In truth, a 'greenhouse' is not a great analogy for our planet. It was only realized in 1907 that real greenhouses work by stopping air currents carrying away heat absorbed from the sunlight. In contrast, the atmosphere stops heat from escaping the surface. It's not a disastrous comparison, because the image of 'greenhouse warming' is useful for describing the general idea, but just be mindful that the analogy can't be taken too far.

One of the first to be intrigued by what role gases might have had on past temperatures was an Irish scientist called John Tyndall, who was based at the Royal Institution in London during the mid- to late 19th century. A popular public speaker, his afternoon lectures used to contend with the latest shows in the West End and would routinely block the streets with carriages containing the great and the good. In 1861, he reported to the Royal Society that he had measured the absorption properties of a suite of gases after he had shone a beam of infrared light through them. Tyndall realized that the main constituents of air – oxygen and nitrogen – were hopeless at absorbing any infrared light and caused virtually no heating. But much to his surprise, Tyndall found that carbon-containing gases in the air – he measured carbon dioxide and ethylene – could play a huge role. He also realized that it wasn't just these gases that might be significant; water and nitrous oxide (better known as laughing gas) also had a big effect. Tyndall suggested that differences in the amount of these gases in the atmosphere might have driven past changes in climate.

The great American scientist Samuel Pierpoint Langley, who was director of the Allegheny Observatory in the late 19th century, followed up Tyndall's work. Langley set up a series of experiments on Mount Whitney in California to measure how much heat the atmosphere absorbed. Seemingly tirelessly, he would routinely trek up the mountainside to make his measurements. On his many visits he noticed 'fossil' trees above the upper limit of today's tree lines and wondered whether past changes in the way heat is transmitted through the atmosphere might have played a role. In 1894, he produced the first measurements of which wavelengths were absorbed at the Earth's surface. Crucially he found that infrared light was absorbed by water and carbon dioxide. The game was on.

It was a Swede called Svante Arrhenius who picked up the baton. Arrhenius' pioneering work in Stockholm put forward the strongest case that carbon dioxide was a major player in past climate changes. He was the first to calculate how much a change in the concentration of carbon dioxide within the atmosphere would affect the temperature. Drawing on work by a geologist called Arvid Högbom, Arrhenius wondered whether a fall in the number of volcanic eruptions might have led to less carbon dioxide in the atmosphere, plunging the world into an ice age. In 1895, he drew on Langley's results and crunched the numbers. He found that a halving of carbon dioxide should bring the world's temperature down by 5 °C. But an increase in carbon dioxide by 50% might lead to a warming of 3 °C and a doubling should drive temperatures up by nearly 6 °C. Arrhenius also realized that as temperatures increased, more water vapour would end up in the atmosphere, trapping more heat and exaggerating any initial warming from carbon dioxide.

But Arrhenius had no comprehension of just how fast we could put greenhouse gases – what he called 'selective absorbers' – into the atmosphere. He thought it would take a few thousand years before the use of fossil fuels would increase the amount of carbon dioxide in the atmosphere by half again. And even if it did, it probably wouldn't be a bad thing: 'we would then have some right', he suggested, 'to indulge in the pleasant belief that our descendants, albeit after many generations, might live under a milder sky and in less barren natural surroundings than our lot at present'. It does sound great.

But not everyone was sold on the idea. One complaint was made by Swede Knut Ångström at the turn of the 20th century; he did some experiments and argued that water vapour blocks much of the same infrared radiation as carbon dioxide; it therefore can't have played much of a role in past climate change. Ångström was essentially saying that the water content of the air was the most important factor; you could change carbon dioxide all you liked, it wouldn't make any significant difference.

Ångström's rebuttal swayed a large part of the scientific community with the result that Arrhenius' idea was pushed to the fringes. A brief attempt in 1938 by the British engineer Guy Callendar to resurrect the concept unfortunately also fell on deaf ears. It was only realized in the 1940s that Ångström was wrong; the absorption properties of greenhouse gases change with altitude, yet Ångström had completed all his experiments at sea level; his results couldn't be extended to the whole atmosphere.

Why Ångström was wrong is actually a crucial point and it's worth just pondering in detail for a moment because it's an excellent way to explain how global warming really works. Imagine that the atmosphere is carved up into lots of horizontal layers from the surface up into space; the lowermost layer contains a mixture of gases, including carbon dioxide and water. When infrared is given off by the Earth's surface, some of it gets intercepted by the gases in this lower layer; these then may give the energy back out or use it to collide with neighbouring molecules, thereby warming the air. As you go up through the atmosphere, each layer does the same with the energy it receives from down below. Ultimately, less infrared reaches the upper levels. This is all fine and dandy when the concentration of gases in the atmosphere stay the same. The amount of energy entering the system is pretty much in equilibrium with what is going back out to space so the temperature remains largely constant. Unfortunately for Ångström (and arguably for all of us), although the wavelengths of energy caught by water and carbon dioxide broadly overlap at sea level, they don't the higher up you go. Big gaps open up. As the altitude changes, water vapour absorbs different wavelengths of energy compared to carbon dioxide. These gaps can be plugged by carbon dioxide, which we're now happily pumping into the atmosphere. The result of this is that with more greenhouse gases in the atmosphere, less heat is escaping to space and the whole place is warming up.

In spite of its uncertain role, the beginning of the 19th century marked the first measurements of carbon dioxide in the air. In contrast to the common gases like nitrogen and oxygen which are measured in terms of percentages, most greenhouse gases are in relatively low abundance. You might have read that the amount of carbon dioxide in today's air is around 380 parts per million – or ppm – and rising. The ppm simply means the proportion of a gas relative to the total amount of other gases in dry air. Because the amount of moisture in the air is so variable across our planet – consider the difference between the Sahara Desert and the Highlands of Scotland – the water content is not included in the calculation for the purposes of comparison. So 380 ppm is the same as writing that there are 380 molecules of carbon dioxide for every million molecules of dry air. There are of course other greenhouse gases, some natural and others produced by humans. Methane is arguably the best known, probably because of its association with flatulent livestock. Significant amounts are also produced by wetlands. Although there is a lot less methane in the air than carbon dioxide, it absorbs somewhere between 20 and 25 times more heat. Good old laughing gas – nitrous oxide – is naturally produced by soils and the oceans. In the past its levels were fortunately low, since nitrous oxide is 300 times more effective than carbon dioxide at heating the atmosphere. Together, they all add up to keeping our planet at the balmy temperature we enjoy today.

The question is, if there was drastic warming 55 million years ago, what were greenhouse gas levels doing at the time?

A whole host of different carbon sources have been suggested for the extraordinary isotopic shift seen during the Palaeocene–Eocene Thermal Maximum. It pointed to an enormous amount of gas going into the atmosphere with very little carbon-13. Comets were one idea. A global conflagration of burning peat and coal has also been proposed. But the very low carbon-13 in the oceans and land smacked of methane. Compared to the other carbon gases, methane has little of the heavier isotope. Not only this, but vast reserves of it exist in the form of ice crystals known as methane hydrates (sometimes called clathrates). Estimates vary, but the amounts involved are enormous; somewhere on the order of several thousand gigatonnes of carbon exist in this form. To put this in some sort of perspective, this is equivalent to all other fossil fuels combined. If just 10% of the methane hydrates were to escape in a few years, it would have the same effect as a tenfold increase in atmospheric carbon dioxide. This is apocalyptic stuff.

Methane hydrates are found within the pore spaces of cold offshore sediments on the continental margin, and in some frozen ground (such as in Siberia). In the oceans, the methane is formed by microbes that break down organic matter after it has fallen to the sea floor, mostly within water depths of 300 metres. The resulting methane has very little carbon-13. If the methane is released relatively slowly from the ocean floor, other microbes can break it down quickly, so that it causes relatively few problems. But if the release is large and from a shallow part of the continental margin, most of the methane could get to the surface unaltered.

Unfortunately, although bound up, methane hydrates are not that stable. If the temperature increases and/or the pressure is reduced, there could be a massive belch as the methane is released from its prison. Methane hydrates have been spotted seeping out off the coast of Santa Barbara, California. It's clearly happening today. The big unknown is whether this is unusual and the start of something bigger.

During the PETM, the evidence certainly points to methane driving the warming. The carbon isotope shift is a strong clue. But so too is the fact that in some of the north and central Atlantic ocean cores there are no foram shells. If gigatonnes of methane were released into the ocean, it would have exhausted the oxygen of the deep sea as it was turned to carbon dioxide. And if that weren't enough, lashings of carbon dioxide would have made the oceans more acidic, dissolving any carbonate. It wouldn't have been a pleasant time to be around. With the higher atmospheric temperatures during the PETM, more warm water would also have got into the deep ocean, causing more methane hydrates to melt. But this sounds a bit like a chicken and an egg situation. Escaping methane may have exacerbated the heating of the ocean, but what drove the first pulse of warming? Something must have started the warming in the first place. The problem is that the vast amounts involved should have left physical scarring on the seabed. But nothing relating to methane hydrates has been found. Could another source of methane be the cause?

Although methane hydrates are the drug of choice for the apocalyptic soothsayers, there is a surprisingly more worrying scenario. In 2004, Henrik Svensen and colleagues at the University of Oslo reported a fascinating find. Within the Møre and Vøring basins of the Norwegian Sea they found a huge sill comprising basaltic rock that had intruded within ancient, organic-rich mudstones. The sill had formed when lava welled up from deep below the surface during the opening up of the Atlantic some 55 million years ago. Svensen and his colleagues found several hundred vents emanating from the sills that had punched through the overlying mudstones. They suggested that when the sill had formed, the carbon in the mudstones would have been baked to temperatures greater than 100–200 °C. The baking would have produced methane that escaped explosively to the ocean floor and beyond via the vents they discovered in the Møre and Vøring basins. Could the shift in the isotopes have been driven by methane formed from baked mud? It would certainly explain the initial massive temperature increase. But it was unclear whether the thick layers of basalt were formed at the same time as the warming began.

Now before we go any further, let's just draw breath and sum up what we think happened during this crazy period. We can split the PETM into three phases. First, it looks like a vast amount of carbon was rapidly released into the ocean and atmosphere. This probably happened in less than 2,000 years, but drove massive warming over 30,000 years or so. After this period, warm conditions persisted for another 60,000 years. It then took another 70,000 years before temperatures started to drop and get back to what they were before. The uncertainty was whether the warming really did happen during the opening of the Atlantic. Was it just a coincidence of similar ages?

In 2007, Michael Storey and colleagues from Roskilde University addressed this question. During such a momentous event as the birth of a new ocean, thousands of volcanic eruptions would have spewed somewhere between five to ten million cubic kilometres of magma. It was one of the largest splurges of magma in the past 250 million years. Storey and colleagues precisely dated the duration of this massive phase of eruptions. Critically, they showed that the Atlantic spreading and warming of the PETM happened bang on the same time. And the age? 55.6 million years ago.

Although methane only stays in the atmosphere for about 12 years before being broken down, it can have a devastating warming effect; not only is it a powerful greenhouse gas in its own right, but the methane gets turned into carbon dioxide which lasts for a further 100 years or so in the atmosphere, continuing the warming effect. But reconstructing past changes in the amount of carbon dioxide is notoriously difficult. On the scale of millions of years, we have to resort to proxies of what was in the air. Anything we can draw on will give at best an indirect measurement and is limited to only some key gases. Fortunately, in spite of what is a relatively small concentration, carbon dioxide does have a major impact on different biological and chemical processes.

One of the better-known proxies for carbon dioxide are the stomata found in leaves. These 'breathing holes' act as the point of exchange for carbon dioxide and water, allowing the plant to grow and respire. It's a delicate balance; if the stomata are open for too long, the plant can die from loss of water; too little and not enough carbon dioxide is fixed by the plant for growth. However, plants are generally more efficient at fixing carbon dioxide when levels are high, and as a result they tend to lose some of their stomata as the concentration goes up. The problem is that different proxies reconstruct different atmospheric carbon dioxide levels during the PETM. Estimates range from 100 to 3,500 ppm, with leaf stomata often giving the lowest reconstructed values. Given what we know was happening to the carbon cycle at the time, the low end of this range seems pretty unlikely.

An alternative approach was recently suggested by Tim Lowenstein and Robert Demicco at the State University of New York. Instead of looking at a biological system to understand what the atmosphere was doing during the PETM, they proposed using the known properties of minerals made up of sodium carbonate.

Sodium carbonate can form a plethora of minerals that precipitate within waters exposed to different concentrations of carbon dioxide. In the western United States, long-lasting lakes laid down sediments rich in minerals of sodium carbonate through the PETM. In the Green River basin, a 300 metre thick section of nahcolite (sodium bicarbonate) has been found: within the nahcolite are small layers of common salt. Experimental studies show that the presence of nahcolite and salt means that the level of carbon dioxide in the air must have been more than 1,125 ppm. With temperature reconstructions for the area, the Green River basin suggests that the early Eocene carbon dioxide levels were somewhere between 1,125 and 2,985 ppm. This is a staggering amount; it's way beyond the level we have today. These values would certainly explain the humungous warming seen during the PETM. But it didn't last forever. Above the Eocene levels in the Green River basin, the main sodium carbonate mineral switched to something known as trona, which can't precipitate above 1,125 parts per million. By around 20 million years ago, all the evidence seems to point to carbon dioxide falling to levels comparable to today.

So it looks like the oceans were the origin of this nightmare period. Yet as the climate shifted to wetter and warmer conditions, productivity in the surface of the ocean boomed. Off what is now New Zealand, this appears to have been driven by an increased supply of nutrients from neighbouring land masses. Over time, the oceans seem to have acted as one enormous biological pump, sucking greenhouse gases out of the air and dumping the organic matter onto the seabed, locking away the carbon. But this might not be the saviour we'd like to imagine for the future. It's not the great get out of jail free card that it might first appear. It took around 70,000 years to get the temperatures back to where they were before the PETM. There's no doubt it can help over the long term, but it doesn't look like it had much immediate effect 55 million years ago.

It's easy to just focus on carbon dioxide levels when we look at global warming, but if we really want to understand the scale of the problem we need to take account of all the other greenhouse gases. The different heating values of these gases can be combined to give a value equivalent to carbon: the so-called carbon dioxide equivalent. How high can the carbon dioxide equivalent get before we're in trouble? Arguably we're already in trouble, but if we want to avoid setting off a series of changes that will be impossible to stop, we need to keep world temperatures from rising more than 2 °C. This might not sound much, but it hides a range of regional climate extremes that would result in devastating impacts on our planet's flora and fauna. How this translates to greenhouse gas levels is still being debated. Essentially it's all expressed as probabilities of risk. A recent paper suggests that the odds are only in our favour of keeping below 2 °C – with an estimated risk of 28% – if the equivalent level is kept to 400 ppm; a value of 550 ppm has a 68 to 99% chance of breaching 2 °C. It's not very heartening when we add together the heating capability of today's greenhouse gases and find we're at around 460 ppm. We're already beyond a safe level.

Although the amount of greenhouse gases in the atmosphere is dangerously high, this is just part of the story when it comes to warming our planet. Other factors also play a role in controlling the temperature of our atmosphere, and some of these help to cool things down. One example is sulphur dioxide given off during volcanic eruptions; this can react with moisture in the air to form a fine mist of sulphuric acid droplets. These aerosols have an impact on the world's climate by directly reflecting heat back from the Sun. But they also attract water in the air to form clouds, which in turn can also reflect heat back out to space. The result is that eruptions can significantly cool the planet; the more sulphur that's emitted, the colder it gets. When the Philippines volcano Mount Pinatubo erupted in 1991, the amount of sulphur that was given off helped cool the world's surface by 0.5 °C during the following year. The same sort of effect kicks in with the burning of coal. As more coal-fired power stations are being used to generate electricity – China alone is opening the equivalent of two coal-fired power stations a week – some of the associated sulphur dioxide forms aerosols, helping to offset the direct warming effect of the greenhouse gases and giving us a brief respite from the worst effects of climate change. All these cooling effects mean that on balance we're at about 375 ppm equivalent. But there's a sting in the tail. Aerosols generally have a shorter lifetime in the atmosphere than most greenhouse gases. As we've seen already, a molecule of carbon dioxide hangs around in the atmosphere for around a century, whereas many aerosols only last for several days before they're stripped out of the air by rainfall. The implication is that, as we start to clean up the world's power stations, the cooling effect from aerosols will lessen, allowing the full warming effect of greenhouse gases to become apparent. We have to get our emissions down and fast. Even a delay of five years could be too long.

This might all sound a bit too apocalyptic. After all we're unlikely to keep going on a reckless high emissions spending spree. Or are we? The PETM shows that if you suddenly release a vast amount of hydrocarbons into the atmosphere you get catastrophic warming. Worryingly, although we reconstructed what the carbon dioxide levels were like through this time, we have to remember that this gas probably originated as methane from the oceans. We can use the shift in isotopic values in the forams to calculate how much methane must have been released. To balance up the numbers, more than 3,000 gigatonnes of carbon must have been released. To put this in some kind of perspective, we have relatively easy access to oil and coal that amount to around 5,000 gigatonnes of carbon. The warming 55 million years ago tells us that we mustn't exhaust all of these fossil fuels. If we do, the only thing it will help will be a property boom in Kamchatka.

But if the PETM demonstrates how warm our world can get, is it possible that the Earth's climate is capable of veering to the other extreme? It might seem inconsequential to know if the world can freeze over, but here lies an important lesson on how climate changes can be greatly exaggerated.

Chapter 2
SNOWBALL

Climate change isn't a new idea. In the 4th century BC, the Greek philosopher Theophrastus believed that the number of spots he counted on the Sun's surface could explain the changing rainfall. During the Age of Enlightenment in the 18th century, gentlemen scholars noticed that many Classical writers described a climate different from their own day; Edward Gibbon, for instance, noted in his 1776 weighty tome *The Decline and Fall of the Roman Empire* that the 3rd century AD must have been cooler, citing the frequently frozen Rhine and Danube rivers in support of his idea. After this, things really gathered pace. By the mid-19th century it was realized that ice ages had been common in our planet's past. The nice cosy image of a stable world disappeared forever.

It's hard to imagine that a world of ice can help us understand future climate. With concerns of global warming it might seem academic to try to understand ice age climates. After all, it is the opposite of what we're anticipating. We'll look a little later at how ice ages were discovered, but for now let's just take it as read that they're a fact and look at how they might be useful for understanding what lies ahead.

Although we often associate ice ages with mammoths and strategically clothed cave people, the frozen spells of our imagination happened relatively recently. For a planet that's 4.6 billion years old, hairy jumbos and cavern dwellers were just yesterday. There was a time, however, when things were far worse. Much earlier, a series of global catastrophes seem to have taken place, each cocooning the world with ice. Seen from space, our planet would have appeared as an enormous iceball. No one is seriously suggesting the world is going to turn into a global winter wonderland in the near future, but this severe chill gives us an idea of how far the world can go in the stakes of extreme climate.

A crucial element of all this is identifying what happened, and when, in our early planet's history. It's an issue fraught with problems. Not least of these is finding parts of the world that preserve the past. The further back you go, the less there is. There are a whole host of reasons why this is the case but the bottom line is the world is a dynamic place. The best you can often hope for is an exposure of rock or mud that somehow escaped the ravages of time; preserving in it what happened at the moment of formation. How to date these records of the past plagued early scientists until it was realized that many contained fossils of ancient plants and animals. Over millions of years, the different fossils preserved a record of evolution and extinction: essentially the story of life on our planet. Although fossils don't give a direct age for a rock, they can be placed in an order relative to one another. The geological scale was born. A plethora of weird and wonderful names were conjured up to chronicle the different times of our planet's history. With the development of scientific dating techniques it's now possible to put ages to these past times. We now know for instance, that the explosion of life recorded by fossils in the Cambrian – after the Roman name for Wales where it was first discovered – began 542 million years ago. But for a lesson in future climate change we can go further back in time to when little life existed: the so-called Neoproterozoic.

Although it's a dreadful mouthful, the Neoproterozoic represents a fascinating era that spans a vast 460 million years. In spite of the huge amount of time involved, it was realized late in the 19th century that some parts of the world preserved evidence of early ice ages. In 1871, James Thomson reported ancient rubble at Port Askaig in Islay, Scotland, that had been laid down by glaciers. In 1891, the Norwegian geologist Hans Henrik Reusch followed this up with the discovery of ridges and mounds containing rock and debris that marked the outer limit of an ancient ice field.

This period of advancing ice was named after the site where it was first discovered. Found along Varangerfjord within the Norwegian Arctic Circle, Reusch called his ice age the Varangian. It might sound a bit like we're in fantasyland but bear with me.

The idea that the world was covered by ice at the dawn of time was really first touted by one of the great scientists and adventurers of the early 20th century: Sir Douglas Mawson. Mawson was a *Boy's Own* hero. In 1911, he led an Australian expedition to Antarctica's George V Land to collect scientific samples, reach the Magnetic South Pole and see at first hand how ice might mould the landscape. The team suffered a series of disasters that would be more fitting on the pages of a Hollywood script. After over-wintering at Cape Denison, Mawson led a three-man team to explore the eastern lands, crossing what are now known as the Mertz and Ninnis glaciers. The conditions were particularly harsh at this time and a combination of crevasses, lack of supplies, vitamin A poisoning (from eating the livers of their sledge dogs) and exhaustion resulted in the deaths of Mawson's less fortunate friends and colleagues, Mertz and Ninnis. Against all the odds Mawson struggled back alone for 24 days and reached Cape Denison to be greeted with smoke on the horizon; his ship had sailed away only hours earlier. Fortunately, a small group had volunteered to stay in case of his return and Mawson was forced to spend another winter in Antarctica. His discoveries and tales of endurance rank alongside those of his contemporaries but were eclipsed by the death of Scott in 1912.

When Mawson wasn't leading heroic expeditions to Antarctica, he spent a large part of his career looking at the early geology of the Earth. As the evidence for ancient ice ages mounted up, Mawson pulled all the data together. In 1949 he recognized over twenty Neoproterozoic sites around the world, stretching from the Arctic to the Equator, and argued for the first time that that this widespread pattern showed that the world had 'experienced its greatest Ice-Age'. This was revolutionary stuff, but Mawson wasn't the only one touting new ideas.

Between 1912 and 1915, a young German scientist called Alfred Wegener published scientific papers and a landmark book called *The Origin of Continents and Oceans* in which he attempted to solve a centuries-old conundrum. In 1596, the Dutchman Abraham Ortelius, and later the Briton Francis Bacon, had spotted how easily South America could snuggle up alongside Africa. Charles Darwin had also been intrigued and speculated 'on land being grouped towards centres near Equator in former periods and then splitting off'. Wegener had looked into this further. He realized that rocks found on one continent matched another; for instance, the geology of the Scottish Highlands was identical to the Appalachian Mountains in North America. He also found fossils of tropical species in the Arctic. On the basis of all this and more, Wegener proposed that the continents ploughed their way through the oceans, changing their location on the surface. By 1922, Wegener had developed his ideas further and was arguing that around 300 million years ago the continents had formed one large landmass – a supercontinent – called Pangaea. According to his argument, Pangaea had subsequently split up and formed the world we see today. The critics were scathing of Wegener, largely because he had no real idea of how the continents could move about on the surface; 'purely fantastic' and 'German pseudo-science' were typical of some opinions given of Wegener's ideas. Mawson certainly didn't believe a word of it and took the location of his ice age sites at face value.

Although flawed at one level, Wegener's notion of drift led to the realization in the 1960s that the continents form plates which float on a buoyant layer in the Earth: this idea of plate tectonics explained how new continents and oceans were created, destroyed or rubbed along uncomfortably together. More importantly for our story, it explained the jigsaw puzzle seen today; since at least 3.8 billion years ago, plates have migrated around the world, rejigging their position relative to one another. Ultimately, if we took the latest in time travelling technology to visit the Neoproterozoic, the world's surface wouldn't look like it does today. Mawson's idea of a world of ice was suddenly in trouble. The simplest explanation was that tropical sites showing ice had once been near the poles and subsequently travelled towards the Equator. If so, the world's climate back then might not have been any different from today; the same sort of trend from tropical heat to icy poles probably existed. But to test this it was necessary to delve into the rocks themselves and look for a faint magnetic signal.

It might seem odd that magnetism can help us understand past climate change but it does so at several different levels. Back in the 15th century, there was an amazing mixture of practical benefits and quack urban myths linked to magnets – in many ways we haven't really moved on today. Magnets had been known about since at least the 7th century BC and were widely used as a navigational aid at sea. But it was only in 1600 that the Englishman William Gilbert undertook one of the first methodological studies of magnets. Although he later went on to become physician to Queen Elizabeth I, he is best remembered for his book *De Magnete*. Written in Latin, *On the Magnet* was a huge success. In it, Gilbert debunked a whole host of tall tales, including the belief that garlic disrupted magnetic fields; in spite of this it remained a flogging offence for over another century for British naval helmsmen to eat garlic. Importantly, Gilbert recognized that magnets had poles and argued the Earth was one giant magnet; if a handheld magnet was freely suspended, he noted it should point towards the ground. We now know the magnetic field is created by the molten iron-rich part of the Earth's core and that a compass will align itself to the magnetic field when it is free to do so. This change in inclination gives a strong clue as to where you are on the Earth's surface; a magnet will lie parallel to the ground at the Equator, but at the poles will dip at a right angle.

When volcanic rocks are thrust onto the Earth's surface or sediments are laid down at the bottom of a lake or on the seabed, any magnetic particles that are present will align themselves to the Earth's field. Most of the time this involves iron, and although the signal is tiny, the orientation of these particles can be measured; essentially they're compasses frozen in time. From this it's possible to work out where on the Earth's surface rocks or sediment were first laid down. It doesn't matter where the land goes afterwards, the signal should be preserved; a memory of a past latitude. By the 1960s, measurements of the Earth's changing magnetic field had made plate tectonics mainstream. It was clear that the continents had roamed over the world's surface. Mawson's intransigence on continental drift had lost a lot of support for the idea of an icy world; it seemed more likely that his tropical sites had been at the poles and later migrated towards the Equator. It seemed farcical to many early critics that the whole world could have been covered in a lot of ice. But not everyone thought it was a mad idea.

Working in the Arctic, British geologist Brian Harland worked on ice deposits that were Neoproterozoic in age. He analyzed the rocks and measured their magnetic properties, providing some early support for the idea of plate tectonics. Through the 1950s and 1960s he concluded that the ice age deposits he was finding in the Arctic were formed at a latitude where the magnetic field was parallel to the ground. They had to have originated from near the tropics. All hell broke loose. The critics countered him. This couldn't be right. The magnetic signal was so faint that any slight problem in the lab could screw up the measurement. And if this wasn't enough, water flowing through rocks and sediments can lay down extra magnetic particles, overprinting the original signal. The measurements made by Harland had to be an artefact; something had happened to the magnetism in his ice age deposits since they were formed. An icy world looked to be one big red herring. Sure there was ice over 540 million years ago but it seemed pretty unlikely it could have reached the tropics. The world's climate just couldn't have done that. The idea seemed doomed.

To prove that glaciers had indeed got down to the tropics, an ice age deposit needs to show an unaltered low-latitude magnetic signal. This is easier said than done. It's difficult to demonstrate that a signal has been preserved over hundreds of millions of years. It took some three decades before the evidence was found in the Flinders Ranges of South Australia. Within the beautiful Pichi Richi Pass is a rather odd looking set of rocks that seem to be covered with parallel lines. It looks like someone has come along and compulsively drawn a great number of strokes on the stone. But this couldn't be further from the truth; these markings reveal the presence of an ancient ice age estuary.

Making up part of what is known as the Elatina Formation, the conspicuous rocks in the Flinders Ranges are packed with information, preserving a record of the changing tide some 635 million years ago. Each day the sea came in and out, and stacked layer upon layer of sand and silt on the estuary floor. By good fortune, these layers have survived at Pichi Richi, and perversely it's because of their great age. Today the sea is a living frenzy, with small creatures thick on the ground, churning up the ocean floor to salvage anything that can be consumed; fortunately, back at the time when the Elatina sediments were being laid down, little life existed to mix up the layers. By measuring the changing thickness of the layers, it's possible to calculate the length of day when the sediments were laid down. 635 million years ago, a day was somewhere around 22 hours long. From similar sediments dating back 900 million years ago, a day was shorter still at 21 hours. The cause of the ever-lengthening day is the increasing distance the Moon is putting between us – today it's nearly 4 centimetres a year – causing the Earth to spin ever slower on its axis.

Critical to our story is that some of the layers in the Elatina Formation contain stones that couldn't have got there from the scenario painted above; the layers are too delicate to have formed within an ocean capable of transporting big stones. They must have been delivered by melting ice passing overhead, dropping any transported rubble into the estuary sediments below. There are quite a few of these dropstones in Elatina, showing that around 635 million years ago there was enough ice about to form glaciers that reached out over the sea. The question is: where on the Earth's surface were the tidal Elatina rocks formed?

Looking at the orientation of the magnetic particles in the rocks seemed to suggest they had been laid down in the tropics. But was this real? It could easily have been the case that the magnetic signal had been overwritten. Joseph Kirschvink at the California Institute of Technology concocted a great test of this challenge. Some of the rocks from Elatina were clearly folded, indicating that after the estuarine sediments had been laid down they'd slumped and then hardened. If the magnetic signal had been overwritten, then all the particles would be parallel to one another, regardless of where they were relative to the layers. But if the particles were aligned parallel to the layers, following the folds and bends, it would show that the signal was authentic and hadn't been changed since the sediments were laid down. Convincingly, when the samples were analyzed, the particles were shown to be parallel to the layers. The estuarine muds and dropstones had been formed very near the tropics. Kirschvink was inspired and in 1992 coined the phrase 'Snowball Earth' to describe an ice-smothered planet. The idea, however, continued to lie on the fringe of acceptability.

Things really only hotted up in 1998 when Harvard University's Paul Hoffman and colleagues described a sequence of rocks in Namibia that appear to have formed around the same time as those in the Flinders Ranges. In a major article published in the journal *Science*, Hoffman's team recognized a sequence of carbonates and glacial debris, capped by yet more carbonate. Here, the exposed rocks appeared to have formed on the edge of a former sea in a setting akin to the Bahamas. Crucially, the magnetic characteristics of the rocks showed that the ice had been around when the site was at 12° S. The debris contained dropstones, similar to those found in the Elatina Formation, but was also made up of lumps of carbonate gouged from older rocks as the glacier had migrated over the land. Just as Mawson had realized that much of the action driven by the Antarctic ice was happening under his feet and out at sea, most of the evidence for the Namibian ice was preserved in what had been an offshore environment. Here was crucial support for the idea that ice had existed in the tropics.

Suddenly a planet of ice seemed possible; the world's climate actually seemed capable of going to such an extreme. Since the work in Namibia, there has been a plethora of studies. At least 16 sites have now been found with evidence for ice, most capped with a thick layer of carbonates many metres in thickness. Importantly, all of them seem to have formed close to the Equator, with none found at latitudes greater than 60°. But it doesn't seem to have happened once; there were at least three immense events between 710 and 580 million years ago. The ice ages preserved at Pichi Richi and Namibia look like they were formed during the same glaciation, known as the Marinoan. But there were at least another two ice ages: the Varangian described earlier and another called the Sturtian.

But this all raises the obvious question: what impact did these enormous upheavals have on early life? We now know that the earliest multicellular life arose before the Cambrian. In 1946, evidence for soft-bodied organisms was discovered in the Ediacara Hills of South Australia; these fossils have been dated to between 635 and 542 million years ago, predating the explosion of live during the Cambrian. That the Cambrian was thought to hold the earliest evidence for life isn't surprising; the hard-shelled organisms of this period were ideal for preservation. The parts of soft-bodied organisms like those in the Ediacaran, however, were rarely preserved in the geological record. Instead, their tracks, recorded on soft sediments, have come down to us as fossils. It's still not clear what their relationship is to the major groups living today; the Ediacarans don't appear to have had any major limbs, and almost certainly absorbed nutrients from the seawater in which they lived.

Crucially, algae and bacteria were common in the rocks before the ice ages. After the Marinoan, Ediacaran life suddenly blossomed. Mawson spotted this early on and was one of the first to suggest that warming after a global ice age might have driven the flowering of life on our planet.

You might reasonably assume a global ice age would snuff out all life. Yet here we are. There must have been some refuges for living things to continue. Life might have conceivably struggled on around hydrothermal vents at the bottom of the sea. Alternatively, perhaps enough light could have penetrated the ice so as to allow photosynthesizers to survive in the oceans. Conversely, some mountains may have been high enough to get above the ice and provide asylum. We know these sort of environments are capable of supporting some forms of life today. We're still not sure what happened during the Marinoan, but it is tantalizing that such extreme conditions might have given life a kickstart.

This all sounds great but let's just think what we're saying here. Sometime in the past the world was almost entirely covered in ice. How could this be? The results from Pichi Richi show that our world was rotating faster than it does today; this would have meant that much of the heat received from the Sun in the tropics would have stayed in the tropics. Although this should mean that ice ages were common at the poles, the tropics should have been relatively warm. There shouldn't have been ice everywhere. What on Earth was going on?

How could a Snowball Earth have come about? At the time, the Sun was 6% less bright than it is today. This would have helped with the cooling, but it immediately throws up another conundrum: how did the Earth ever escape the ice? To all intents and purposes, once a planet becomes fully frosted, it should stay as such. Yet manifestly this is not the case. The whole thing seemed to be a big puzzle.

So what could have caused a Snowball Earth? One possibility that was touted early on was that the Earth may have rotated at a different angle compared with today. This might sound a bit bizarre but it's a thought-provoking idea. If you visit a map shop or department store, you'll often see globes for sale, ranging from the cheap and cheerful through to the expensive, heavy-duty affairs. The one thing they should all have in common is that the axis of the globe is set at an angle of 23.5° from the vertical (Figure 2.1). It's this angle that gives us the seasons. When one hemisphere points towards the Sun, it's summertime; six months later, the other hemisphere takes pole position. We'll see why this feature of the Earth is important in explaining more recent ice ages, but for the moment let's just accept that we know that the angle can change within a few degrees.

If you've got an old battered globe packed away in the kids' cupboards or suddenly get the urge to buy something more presidential, try a little experiment. Get a torch and shine the light beam on different parts of the surface, making sure the torch is level with the floor. You should notice that in the tropics, the beam of light is tightly focused on one spot. When you switch to the poles, the light should 'spread' out on the surface, reducing the amount of light (and heat) that falls on one area. Now if you increase the angle of

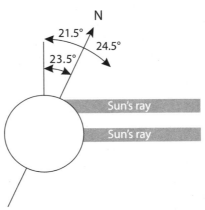

Figure 2.1 The tilt of the Earth.

the globe slightly and repeat the exercise you'll notice little change in the tropics; the light stays tightly concentrated on one spot. Importantly though, over the poles the torch beam should become less spread out and begin to focus on a smaller spot. A similar effect can be felt when you go outside with the Sun low in the sky: because of the angle, the Sun's rays are spread out so you don't feel much heat; when the Sun moves overhead, the rays become more concentrated and it gets warmer. The practical upshot is that the high latitudes receive more heat from the Sun as the angle increases; later on we'll see that this plays a rather important role in reducing the temperature difference across the globe.

But there's no way that a shift of a few degrees could explain all the ice seen in the tropics more than 580 million years ago. One possibility is that around 4.5 billion years ago, something large struck the Earth, creating the Moon. Such a large impact might conceivably have also made the tilt in the Earth's spin axis much larger than it is today; possibly reaching a massive 54° or more. It's hugely contentious that the Earth was at such a large angle but let's just assume for a moment that this happened. What would be the effect of such a gargantuan angle? If you tilt your globe at home and repeat the exercise, things go a bit funny: the tropics receive a lot less heat from the Sun than the poles. The result of this is that the low latitudes would be in a constant ice age, while the poles would become positively temperate.

One type of rock provides an excellent test of whether the tilt was so drastically different from today: evaporites. These are deposits of salt formed when a lake of saline water evaporates. If it's hot and dry enough, thick layers of salt can form. Today evaporites are found on the tropical side of 30° latitude; few, if any, evaporites form over the Equator because the associated high rainfall instantly dissolves any salt that does form. A study reported by David Evans at Yale University looked at the magnetic signal in ancient evaporites that were formed over the past 2.5 billion years. Importantly, he found that the evaporites were consistently laid down between 10 and 35°, exactly where we'd expect to find them today. It means that when the Earth wasn't experiencing raging blizzards, the low latitudes were hot and dry. The angle of the Earth must have been similar to today.

If this is the case then we're left with the inevitable conclusion that temperatures had dropped enough for glaciers to be found in the tropics; average surface temperatures may have got as low as −50 °C. When Joseph Kirschvink first coined the term Snowball Earth in 1992 he also put forward some ideas as to how an icy world might end. Critically it all comes down to feedbacks. We'll read a lot about feedbacks through this book; they crop up in climate change all the time, exaggerating some aspect of the planet's climate and making changes either larger or smaller. The proportion of the Sun's radiation that's reflected off the surface – the albedo – can be a major climate feedback. An extreme version of albedo is when you're dazzled by light reflecting off any bright object. Freshly fallen snow has a high albedo and can reflect up to 90% of the radiation that falls on it; that which isn't reflected is absorbed and heats the surface.

Under early Snowball Earth conditions, Kirschvink envisaged high albedo as a positive feedback: any ice cover would reflect heat from the Sun, helping cool the planet. As the Earth got colder, the icy areas became larger, reflecting ever more sunlight back into space. Calculations suggested that if more than half of the world's surface was covered in ice you'd get a runaway positive feedback, driven by this ice-albedo effect: the world would become a Snowball Earth.

This positive feedback would have been helped by the distribution of the continents at this time. We know the continents were concentrated over the low and mid-latitudes – forming a super-continent called Rodinia (Figure 2.2) – and this would have added to the albedo effect. Today we find that more heat is received from the Sun than lost to space within the latitudes of 37° north and south; polewards of these latitudes, more heat is lost through reflectance and scattering as the Sun's rays penetrate our atmosphere. If none of this extra heat in the tropics were

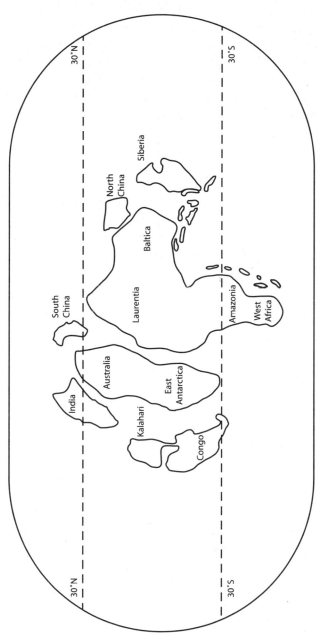

Figure 2.2 The continent of Rodinia 750 million years ago.

moved to the north and south, the temperature in the polar regions would plunge more than 20 °C while the tropics would become about 10 °C warmer. The critical thing here is that the distribution of ice, land and ocean over the surface has a big impact on how much heat is taken up by the Earth in the first place. Although land has a lower albedo than ice (it can be any-where from 10 to 40%), it's a lot higher than the oceans (which can get as low as 4%). The practical upshot of this is that if there is proportionately more land in the tropics than ocean, a greater amount of the Sun's radiation would be reflected. If we butchered our globe and covered a large area of the low and mid-latitudes in grey coloured paper (to represent Rodinia), we'd find that a lot of the light from our torch would be reflected; a lot less heat would be taken up by the planet. It would have all helped develop a Snowball Earth.

But this can't have been the full story. Snowball Earths had to end. But how? The most obvious candidate is warming driven by greenhouse gases, most probably carbon dioxide. If the levels built up high enough, there would be enough heat in the atmosphere to melt our planet's icy shell. A likely source is volcanic activity. The amounts involved are hard to quantify this far back, but today's volcanoes contribute somewhere between 0.1 and 0.3 gigatonnes of carbon to the atmosphere each year; although this is less than 1% of what we're putting into the atmosphere, it would add up over millions of years. To finish off a Snowball Earth, carbon diox-ide in the air would have had to have reached around 350 times modern levels. This is devastatingly high; it would mean carbon dioxide was somewhere around 120,000 ppm compared to today's concentration of 380 ppm and rising.

Today, carbon dioxide is naturally stripped from the air. This can happen in a number of different ways. The world's oceans are one good absorber of carbon dioxide. Perhaps surprisingly, so too are mountains. Carbon dioxide reacts with moisture in the air to form carbonic acid, and this can eat away at carbonate and silicate rocks; the more mountains there are, the more rocks are exposed and the more carbon dioxide is taken out of the atmosphere. If the sea and land were covered in snow and ice, these natural sinks for carbon dioxide would have effectively stopped working, allowing greenhouse gas levels to build up in the air. At first, a runaway ice-albedo would have kept the Earth in a near-permanent winter wonderland. But after around 10 million years or so, the carbon dioxide levels would have got high enough to override the cold. Once the ice had started to melt, a landscape of smashed rock would have suddenly been exposed to the elements. There would have been an all-out assault, with massive amounts of rock being broken down by a rain of carbonic acid, making the acid rain of the 1970s seem like a Sunday picnic. The broken down carbonate rock would have been washed into the world's oceans, making them extremely alkaline and forming the thick cap carbonates seen in Namibia and other Snowball Earth sites.

The critical point that Kirschvink made was that, given enough time, the carbon dioxide levels in the air would have become large enough to override the ice-albedo and ultimately shift temperatures the other way. Once the temperatures had become high enough, the snow would begin to melt and the albedo would crash. Wet snow has a far lower albedo than the freshly fallen stuff (it's about 40%), allowing more heat to be absorbed by the surface. The snow would melt and the exposed land would shift to an albedo as low as 10%, helping drive an enhanced greenhouse effect.

Needless to say, not everyone is convinced. Some researchers have argued that it's more likely the tropical oceans were actually ice free when the continents were experiencing surging glaciers. This has been likened to one enormous slushball, where the open oceans would have allowed some form of life to continue. The flip side of all this is that with large expanses of open water it would have been hard to keep a slushball going as long as a Snowball Earth. Because more ocean would have been exposed to sunlight, the planet could have responded more quickly to the carbon dioxide gas being belched out by the world's volcanoes. Given this, a slushball should have only lasted around a million or so years.

It's possible to test this by using an exotic element only found in abundance within meteorites and deep in the Earth: iridium. Bernd Bodiselitsch and colleagues of the University of Vienna looked at several cores of sedimentary rock spanning the end of the Marinoan Snowball Earth. They reasoned that if iridium is constantly being showered onto the Earth's surface as meteorites burn up in the upper atmosphere, most of this element should have been locked up in the snow and ice of a Snowball Earth. With warming, however, the snow and ice would melt, flushing all the iridium on the surface into the oceans to become locked up in the sediments as a distinct layer. The longer the icy conditions lasted, the more iridium should be present in the spike. By measuring the amount of iridium through the cores, Bodiselitsch's team found a huge spike of iridum within the sediments at the end of the ice age. Other elements within the level of the spike showed that meteorites were the most likely source of the iridium; volcanic eruptions have a distinctly different suite of elements. Using the known rate of meteorite strikes on Earth over the last 80 million years, the amount of iridium falling through from the sky could be used to calculate how long the ice had covered the surface. The iridium in the spikes suggests that the Marinoan lasted around 12 million years; far too long for a slushball. It looks like snowballs ruled.

We know that there were at least three ice ages through the Neoproterozoic, suggesting that the climate went from extreme warmth to extreme cold and back again. The most likely explanation is that once much or all of the snow and ice had melted, the high albedo brought on by the low latitudes of the continents would have started to reflect the sunlight off the planet. The high levels of erosion caused by the presence of continents in the tropics would have helped bring the carbon dioxide levels back down from their mega levels. All these factors would have conspired to kickstart yet another Snowball Earth. The Earth only seems to have got off this extreme climate helter skelter when Rodinia broke up and some of the continents migrated to higher latitudes, calming everything down.

The conditions during Snowball Earth show just how far our planet can go when feedbacks start to kick in. Fortunately, a return to these extreme cycles seems pretty unlikely in the near future; the continents are unlikely to start converging on the tropics any day soon. The critical thing is that the combination of greenhouse gases and albedo play a major role in controlling the temperature of our planet. As we've just seen, albedo can have a massive effect on the amount of energy our planet reflects and absorbs from the Sun, while greenhouse gases played a pivotal role in ramping up the temperature enough to break the icy dead-lock of a Snowball Earth. It's a sobering thought that although carbon dioxide is naturally stripped from the air by the weathering of rocks, it's not likely to help us much today; it's been estimated that this process would take around 80,000 years to get the amount of carbon dioxide in the air down from 500 to 400 ppm. But could these sorts of feedbacks conspire against us in the future? To answer this we need to start looking at the more recent past.

Chapter 3
A BIT OF A CHILL

We have to be a little cautious when taking lessons from the dim and distant past. Although the previous examples of past climate change give us important insights into what the future portends, we must not take them too far. There is little doubt that greenhouse gas levels drove the warming 55 million years ago, but the world's sea level was a couple of hundred metres higher than today. Even if all the ice in the world melted and the ocean waters expanded as they warmed, a sea level of this magnitude is unlikely to happen in the future; it was only possible in the past because the rate at which the tectonic plates were tearing themselves apart pushed up the sea floor. If we want to be on more solid ground when extrapolating from the past, we need to fast forward in time. We need to go to a period when the patchwork of continents was pretty much the same as today.

It's clear that since the hothouse conditions of 55 million years ago the world has become a cooler place. Why is one of the big questions. As always there are a host of likely scenarios. One possibility is that when Antarctica separated from South America sometime between 25 and 20 million years ago, it allowed the development of the Antarctic Circumpolar Current. This enormous current sweeps around Antarctica, linking up the Atlantic, Indian and Pacific oceans and effectively isolating much of the continent from the rest of the world; this could conceivably have helped foster the growth of a major ice sheet. But this isn't the only contender. If you fancy another notion, North and South America joined together around 3 million years ago, closing the Isthmus of Panama. This would have stopped warm tropical

Atlantic waters flowing into the Pacific, the result of which would have been to increase the saltiness of the Atlantic Ocean, allowing deepwater formation to take place and stopping warm tropical waters reaching into the Arctic and keeping it ice free. One of the most popular solutions is that as India slammed into Asia around 50 million years ago, the Tibetan Plateau rapidly began to form. As this enormous block was thrust upwards, it would have exposed increasing amounts of rock, effectively acting as a sponge, stripping the air of carbon dioxide. Recent work now shows that the first ice began to develop in the Arctic around 45 million years ago, about the same time as Antarctica, suggesting that a drop in carbon dioxide may have been the cause of increasing ice cover in both hemispheres.

Regardless of the cause, the result was that by 2.6 million years ago the Earth had become locked into a bout of ice ages that continue to this day; a period known as the Quaternary. Although these ice ages weren't of the same scale as Snowball Earths, they were still pretty big. It was a time when vast ice sheets reached out from the poles to the mid-latitudes and then shrank back; when our ancestors migrated out of Africa to colonize the world; and when we moved away from mammoth steaks to all-you-can-eat restaurants. Crucially, it's the regular and frequent changes in ice through this period that gives us a solid basis for understanding long-term climate change.

It was back in 1837 that a youthful president of the Swiss Society of Natural Sciences called Louis Agassiz first coined the expression 'ice age'. Agassiz was an expert in fossil fishes, but became exasperated by colleagues who suggested glaciers were a lot larger in the recent geological past. Convinced that he could prove otherwise he soon became a convert and then went a whole lot further, touting the idea of a world of ice. He got thoroughly carried away with the evidence and seemed to see an ice age everywhere. He visited Brazil in the 1860s where he found what are now known to be heavily weathered tropical soils. Agassiz believed they were the result of a Great Ice Age and proclaimed he had found the final evidence of an icebound world. He envisaged 'the development of these huge ice sheets must have led to the destruction of all organic life at the Earth's surface'. It was catastrophic stuff. Agassiz took the view that the ice ages were divinely inspired; all life was wiped out in one big snow blitz so that it could start anew again. Unfortunately, as a result of these creationist beliefs, Agassiz couldn't see the merit of Charles Darwin's theories on evolution and remained bitterly opposed to them to the end of his days.

In spite of Agassiz's over-the-top exuberance, it was clear he wasn't totally mad when it came to the higher latitudes. Some of the earliest support came from Britain. With this new fangled idea of an ice age, the landscape started to make some sort of sense. The debris littering the surface was once thought to be from the biblical Great Flood. Now it could be explained by something far more tangible. Piles of rock weren't formed when Noah was dodging the waves; they must have been created in the presence of an ancient ice sheet.

We now know there was more than one ice age in the past; at the last count it looks like there were at least 50 of them through the Quaternary. The last one peaked around 21,000 years ago and is often referred to as the Last Glacial Maximum. At this time, most of the British Isles was covered by one large ice sheet. All of Scotland was subsumed, along with most of Ireland, Wales and northern England. The ice, however, didn't just sit there. Just as a blob of ice cream will flow away from the highest point, a slab of ice will also move if given enough time. As snow builds up on the upper slopes of an ice sheet or glacier, it will eventually turn to ice and flow downhill. In itself, this sort of flow isn't terribly fast. But ice can also melt under pressure, even at subzero temperatures; this can be driven by heat coming up from the Earth's interior. The result is that many ice sheets have a layer of water at their bottom, allowing the ice to slide that much faster than if it were left to gravity alone. If this happens, streams of ice move out from the domed high points, draining the ice sheet and travelling up to 1000 metres a year. The British Ice Sheet looks like it had several high points of ice; although their precise location isn't known, they were almost certainly centred over the more mountainous parts of the country.

Ice streams are essentially rivers of ice. And just like rivers, they can be laden with debris. Ice isn't always the clean translucent thing most of us get from the freezer for our gin and tonics. When it flows over the landscape, it can pick up anything that isn't fixed to the ground. Gravel, pebbles, boulders; it really doesn't matter. Sometimes this material gets dragged along at the bottom of the ice, scarring the surface below and showing the direction in which the ice had flowed. When Agassiz tried to convince some of his Swiss colleagues of this, he showed them parallel grooves gorged into polished mountain rocks; his sceptical companions retorted that a horse-drawn carriage most likely formed them. There's no pleasing some critics.

On land, when ice streams reach the edge of an ice sheet, they dump the rock and debris they've picked up, forming vast ramparts known as moraines. At the coast, the ice may float out onto the sea to form a shelf from which great cliffs regularly collapse, releasing icebergs into the ocean. It's the piles of debris that form on the edge of an ice sheet that mark their outer limits, and these can be used to reconstruct their size. It's from this that we know the British Ice Sheet was up to 1.5 kilometres thick and had a volume approaching 800,000 cubic kilometres. Britain would have seemed like one enormous icebox.

Although this all sounds rather hellish, the British Ice Sheet was a pretty small affair. It was one-third the size of today's Greenland Ice Sheet and a twentieth of today's ice over Antarctica. It was a relative minnow compared to the main players; fully melted it would have raised the world's sea level by less than a metre. The neighbouring ice over Scandinavia was more than 2 kilometres thick and had a volume of around 6 million cubic kilometres. The vast North American Laurentide Ice Sheet was larger still: a staggering 3 kilometres thick at its highest point with a volume of 33 million cubic kilometres; it was so large it physically changed the way weather systems moved over North America. In the southern hemisphere, there really was only one main body of ice and this was on Antarctica; although it was probably not much larger than today, it was comparable in size to the Laurentide Ice Sheet. All in all, nearly a third of the world's surface was covered in ice. A huge amount of freshwater was locked up on land, and as we'll see shortly, this had a big effect on the world's climate at this time.

Outside the confines of permanent ice cover, there would have been little respite from the cold. Southern parts of Ireland, Wales and England may have been beyond the limits of the ice sheet but the freezing conditions would have been a far cry from the balmy state we know today. Even when the summers defrosted the surface, the soil would have remained permanently frozen a metre or so below. In these permafrost regions even rocks weren't safe. Temperatures were so low that the surface would have been strewn with shattered stone. It's not the direct effect of the cold as such; it's what happens to the water that's seeped into cracks that counts.

When water freezes, the volume expands by around 9% as ice crystals form. Traditionally it was argued that repeated expansion, thawing and expansion should wedge open rocks and cause them to break up. A good example of this is when you've gone to quickly chill a bottle of wine in the freezer and forgotten all about it; the next day you're invariably greeted with a cracked or shattered bottle accompanied by a block of frozen Chardonnay. The bottle has a fixed volume, so the only thing left for the growing ice to do is to distribute itself over the peas and chips. Recent work by Julian Murton at the University of Sussex and colleagues showed that although ice-shattering bottles of wine might sound like a good analogue for frozen ground, it's unlikely to be true of the real world. For the same thing to happen in permafrost environments, the rocks have to be completely submerged in water, something that is not common in areas where shattering takes place.

Although permafrost regions are largely frozen under the surface, some water still exists as liquid. In spite of the sub-freezing temperatures, this water flows towards where it is frozen in the ground to form lenses of ice a few millimetres thick. As these lenses grow, the ground heaves. Murton and colleagues wondered whether this might also happen in porous rocks as well. Taking a block of limestone, they kept the lower half permanently frozen. The top half of the rock was then heated and frozen 20 times to simulate 20 years in the permafrost. What they found was fascinating. Enigmatically, it was during the 'summertime', when the upper part of the block was defrosted, that most of the ice lenses formed; the newly melted water in the upper part moved towards the lower frozen block, forming ice lenses that broke up the rock.

Yet in spite of the frigid conditions during the last ice age, there wasn't a return to Snowball Earth. The continents weren't concentrated over the tropics and reflecting most of the Sun's rays back out to space. Although there was a glut of ice near the poles, the rest of the world was not frozen over. In spite of Agassiz's early protestations, the Amazon was not bulldozed by ice. In tropical South America, it looks like it became wetter in the southern part and drier to the north. On the other side of the Pacific, conditions were equally challenging.

In some areas of Australia, it does appear to have been freezing during the last ice age. Tim Barrows and colleagues at the Australian National University have dated frost-shattered boulders in the more mountainous parts of southeast Australia. They found that these features formed at the same time as the last ice age. For this to happen, temperatures must have been 9–11 °C cooler than now. The temperatures look like they were low enough for small glaciers to exist in the alpine valleys of Tasmania and the Snowy Mountains of southeast Australia. But the idea of calling this period an ice age in Oz may seem rather odd. There wasn't ice everywhere. Although it was cold, ice sheets weren't roaring over Australia as Mawson had proposed for hundreds of millions of years ago. Instead, it seems to have been a heck of a lot drier.

If you have a look at a weather map for Australia during the southern hemisphere winter you'll invariably find a big letter 'H' slapped bang over the continent. This high-pressure system is a result of dry, sinking air. As the air rises over the tropics it heads polewards, where it gradually cools and sinks at around 30° latitude. Back in the last ice age, this high-pressure system seems to have been a permanent feature over Australia for much of the year. In the southern hemisphere, air comes down from on high in an anticlockwise direction. On a good topographic map of Australia you can actually see physical evidence for this: ancient dune fields from the last ice age are preserved across the continent making a huge anticlockwise whorl; a relict of an ancient pressure system, physically etched on the landscape. It wasn't ice so much as drought that was the plight of Australia at the time.

It wasn't long after Agassiz had first proposed a Great Ice Age that others joined the race for immortality. How could big changes in the volume of ice come about? What was the driver? The first to have a try was the great French mathematician Joseph Adhémar. In his book *Revolutions of the Sea*, which was published in 1842, Adhémar proposed that the driver of the ice age was the way the Earth orbits the Sun. Adhémar argued that this governed the amount of insolation our planet receives. To get our heads around Adhémar's idea we need to understand how the Earth orbits the Sun and how this can play such a significant role.

It was in 1605 that the German astronomer Johannes Kepler first realized that the Earth went round the Sun in an egg-shaped orbit and not as a perfect circle. Not only this, but the Sun is slightly off-centre. The upshot of all this is that the northern hemisphere spends several days longer in the 'summer' half of the orbit than the winter half. Adhémar argued that this meant that

Antarctica had more dark winter nights and therefore must be getting progressively colder because it receives less heat each year.

But to get an ice age, Adhémar suggested it was another factor that played the crucial role. Because of the gravitational pull on the Equator by the Sun, the Moon, and the other planets, the Earth's rotation 'wobbles'. This might sound a bit surreal, but imagine the axis of the Earth extending out from the North and South Poles into distant space. Over 21,000 years, the wobble causes the axis to trace out a cone, progressively pointing to a different spot in space. After 21,000 years, the Earth's axis returns to pointing at the same point in the sky. The principle is just the same as a child's gyroscope or spinning top – albeit on a very different time-scale.

All this wobbling has a big impact on the timing of the seasons because it changes the way the Earth is orientated to the Sun. The orbital position of the seasons relative to the Sun shifts, and this is called the 'precession of the equinoxes' (Figure 3.1). Because of the fact that the Sun is slightly off-centre, Adhémar knew that during the northern hemisphere summer, the Earth

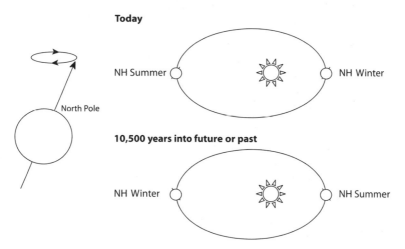

Figure 3.1 The precession of the equinoxes.

was pointing towards the Sun at its furthest point from our star. As a result, he reasoned that halfway through a precessional cycle the opposite would be true; the northern hemisphere summer would gradually move around the Sun until the Earth was the closest it could be. It wasn't a big leap of faith to then suggest that ice ages must happen at different times in different hemispheres. Adhémar envisaged that the polar ice sheets would grow to enormous dimensions, reaching some 100 kilometres in height and resting on the ocean floor. Once precession had warmed things up, the base of the ice sheet would melt and eventually collapse, setting off a devastating series of tsunamis that would destroy everything in their path. It was very much a catastrophic end and in the finest traditions of Agassiz.

It was a brave attempt, but hopelessly wrong. By 1852 the critics rounded on him. The great Prussian scientist and explorer Alexander von Humboldt showed that this just couldn't be right. Although Adhémar was correct that winters would get cold by this mechanism, it was also true that there would be an equally warm summer. Ultimately, the total amount of heat received from the Sun was the same throughout the year. The orbital changes suggested by Adhémar couldn't explain an ice age. But Adhémar was right about one thing, however. During the 1860s and 70s, geologists started finding organic layers in-between glacial moraines across Scotland and North America. This showed that there had been at least one warm period – an interglacial – sandwiched between two ice ages. There must have been more than one 'Great Ice Age'. Could the changes in orbit have been responsible in another way?

It was shortly after this time that one of the great characters in our story entered the fray: the British scientist James Croll. He was an astounding character. In the first four decades of his life he'd been in and out of work and was hardly a model employee. In this time, Croll had been a millwright (though he was apparently not very good), a house joiner (until sore joints stopped him in this pursuit), a tea merchant (this ended when his elbow joint ossified), a temperance hotel manager (which failed because of a lack of customers), an insurance salesman (which he hated because it meant he didn't get much time to think about things) and a newspaperman for a temperance weekly (not a great pursuit for someone who liked his quiet time).

In spite of his rather haphazard employee record, Croll success-fully applied for the post of janitor at the Glasgow Andersonian College and Museum in 1859. He was 38 years old and desperate to get access to the library to develop his own ideas on life, the universe and everything. By 1864, the janitor had published his first paper on ice ages. Croll argued that Adhémar had got it wrong. It was the changes in the shape of the Earth's orbit – its eccentricity – that mattered (Figure 3.2). The gradual change from an elliptical orbit to nearly circular and back to elliptical over 100,000 years was the key. Importantly, Croll didn't believe that ice ages were driven by the total amount of heat the Earth received over a year; instead, he argued, it was the way the heat was distributed over the seasons that mattered.

When the Earth goes around the Sun in an exaggerated egg-shaped orbit, our planet receives more heat in one season than another. The result is that when the Earth is furthest from the Sun, it can have exceptionally chilly winters. During a highly elliptical orbit, Croll argued that major snowfields could build up with a suc-cession of cold winters. He was one of the first to describe climate feedbacks. For ice ages, Croll envisaged a positive feedback: as the snowfields built up, the increasing area of high albedo would reflect what little radiation was making it to the surface; this would drive the temperatures lower and help make the snowfields larger still, eventually resulting in a full blown ice age.

Today, the Earth is travelling around the Sun in an orbit that is more circular in shape. Croll argued that it didn't matter two hoots what precession was doing at the moment; the orbit wasn't nearly eccentric enough to build up the ice. Precession was only important when an egg-shaped orbit was dominant. When it was, Croll agreed with Adhémar that the ice ages must happen in the hemispheres at different times.

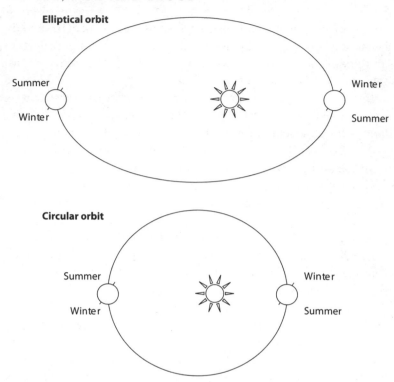

Figure 3.2 The eccentric orbit of the Earth.

But this wasn't all. In 1875, Croll introduced the third and final astronomical feature into the orbital forcing of the ice ages: the tilt or 'obliquity' of the planet. We saw earlier that the Earth can nod backwards and forwards. By the turn of the 19th century it was known that the planet could swing from 21.5° to 24.5° and back again over 41,000 years. Croll suggested that when our planet is tilted at a greater angle, ice ages are less likely; this was because the poles receive more heat through the year.

At the risk of digressing, this change in the Earth's tilt still impacts on us today. At the moment, the angle of tilt is actually becoming less by around half a second of a degree a year. This might not sound much, but even today such a small change has consequences for budding megalomaniacs of the world. A wonderful example is in Taiwan, where in 1908 the then Japanese government decided to commemorate the completion of a north–south island railway line by building a monument on what was then the Tropic of Cancer. It's important to realize that the Tropics of Cancer and Capricorn mark the highest latitude where the noon summer Sun appears overhead; because the angle of tilt is reducing so too are the latitudes of the tropics. As a consequence, the Taiwanese monument is now over 1.3 km north of today's Tropic of Cancer; the tropical boundary will continue to move south for 90 kilometres over several thousand more years before it finally starts heading north again.

When all these factors were taken into account, Croll proposed there couldn't have been an ice age for at least 80,000 years; since this time, the northern hemisphere had been in an interglacial. The problem was that all the best dated evidence at the time suggested the last ice age ended around 10,000 years ago. An orbital origin for the ice ages still didn't seem to work that well.

As a result, not everyone was convinced that Croll had got it right. Arrhenius certainly didn't believe much of what Croll had to say on the matter; he felt that the Brit's proposals were all wrong: carbon dioxide was the driver. But in 1876, geologist Joseph Murphy argued that Croll was mostly right but he'd made one small mistake. It wasn't how cold the winters got that mattered, it was the summers that were the key. Murphy argued that it was low summer warmth that allowed snow to survive through the year that was important. Only when the maximum temperatures stayed low could ice persist and build up into an ice age. This was the opposite to Adhémar and Croll, who had both insisted that extremely cold winters were needed to start an ice age.

In spite of Murphy's suggestion, the idea seemed to fall by the wayside. It was really the energy of another man, a Serbian called Milutin Milankovitch who seemed to provide the solution. He spent a good part of the First World War reworking Croll's ideas.

By 1920, Milankovitch had calculated the combined effects of eccentricity (the 100,000-year cycle), tilt (the 41,000-year cycle) and the precession of the equinoxes (the 21,000-year cycle). He worked out the amount of radiation falling on different latitudes over the past million years. Thanks to his calculations, Milankovitch could argue that it was the land at high latitude, in particular 65° N, that was crucial: this was where the biggest changes in the amount of solar heating took place. Whereas Murphy had failed to convince his colleagues, Milankovitch had the numbers. Taking the argument that cool summers were the key, the end of the last ice age was predicted to end 10,000 years ago, in full agreement with the geological record. Suddenly there seemed to be a working model. Instead of Murphy's Law, the idea became known as Milankovitch's Orbital Theory.

But could the orbital theory explain earlier ice ages? Because ice sheets tend to bulldoze the remains of previous advances, only snippets of the past have survived; the odd bit of a glacial moraine here, an organic lens there. It's hard to get a full picture from these small fragments because they're just a few pieces of a very large jigsaw. How many ice ages had there been before the Last Glacial Maximum and when did they happen? A sensitive recorder of the changing amounts of ice in the world was needed. There didn't seem to be an obvious way forward.

Surprisingly, it's only been since the 1930s that large research ships have ploughed the waves and taken deep-sea cores from the ocean floor. Before this time it didn't seem worth bothering; it was thought the oceans hadn't changed that much in the past. This view took a battering, however, in 1955 when an Italian called Cesare Emiliani reported the results of some work he had done on the shells of foraminifera that were laid down on the tropical sea floor over the past few hundred thousand years.

Just as carbon has different isotopes, so too does oxygen. In this case it's oxygen-16 and oxygen-18. The principle is exactly the same as before: one is heavier than the other but both behave the same chemically. Importantly, the forams take carbon and oxygen directly from the seawater to build their carbonate shells. The important thing as far as Emiliani was concerned is that when it gets warmer, more of the lighter oxygen-16 is fixed in the foram shell; when it gets cooler, more of the heavy oxygen-18 is taken up. He believed that by measuring the different proportions of oxygen isotopes in the forams, it would be possible to get a better understanding of past climate. The results were exciting. Emiliani's reconstruction implied that tropical ocean temperatures had constantly changed over time, and crucially, had done so around the time predicted by summer changes in sunlight reaching the high latitudes of the northern hemisphere. Here was strong support for Murphy's and Milankovitch's idea.

But there was a potential problem: did the oxygen isotopes in the forams really record temperature? Although modern studies showed that changing temperature affected the oxygen isotopes, it wasn't so clear further back in time. Temperature must have had an effect, but was it the only thing that had an influence?

When the temperature drops over the ocean's surface there's a lot less evaporation; what little happens preferentially takes up molecules of water containing the lighter oxygen. The result is that, over time, the heavier oxygen ends up becoming concentrated in the ocean water. In contrast, a significant portion of the evaporated water will eventually fall as snow over the high latitudes, locking the lighter oxygen-16 up in the ice sheets. All this means that during an ice age there tends to be more oxygen-18 in the ocean and more oxygen-16 in the ice. But when the climate warms during an interglacial, the opposite happens. More evaporation takes place, pumping more water with heavy oxygen into the air while most of the ice melts, returning the lighter signature to the sea. Could changes in the amount of ice have played a role in Emiliani's reconstruction?

In 1967, British scientist Nick Shackleton suggested that the benthic forams living in the bottom of the tropical sea were unlikely to have experienced much of a change in temperature. By picking out both benthic and surface-living forams, Shackleton showed that both types pretty much mirrored each other in their oxygen isotope content. If this was the case, then the isotopic makeup of the surface forams can't have been driven by temperature. Global ice volume must have been the cause. Temperature seemed to explain only about a third of the signal; the rest had to be ice volume.

Critically, if we want to understand what the ice did in the past it's not much use looking at records from the ocean floor near the poles; the temperature signal here is too strong and dominates the oxygen isotope values. Fortunately, the ocean circulates as one big conveyor belt around the world. As a result, when ice melts, the oxygen isotope signal is injected into the oceans where it's transmitted around the world and taken up by any forams who happen to be about at the time. Paradoxically, it was realized that to reconstruct past ice volume – most of which was at or near the poles – you needed to look at ocean cores from the tropics.

In 1976, working with Americans John Hays and John Imbrie, Shackleton looked at oxygen isotopes preserved in foram shells spanning the past 450,000 years. Analysis of the record showed there was a series of different cycles embedded one on top of the other. Small but frequent changes in ice volume seemed to happen alongside big, long-term changes. It was like a mad band, each loosely playing the same tune but to their own beat. Critically, when the reconstruction was statistically analyzed it showed the same peaks in ice volume that were predicted by changes in the Earth's orbit: peaks and troughs in ice volume happened every 21,000, 41,000 and 100,000 years. Here at last was the definitive proof that the orbital theory first touted by Adhémar and later developed by Croll, Murphy and Milankovitch was correct. The way the Earth orbited the Sun clearly affected the climate; ice ages happened when the high latitudes of the northern hemisphere got little summer Sun. The changing orbit was the driver of the ice.

But could these orbital changes have been enough in themselves to build massive ice sheets up to 3 kilometres thick? The changes in the amount of heat from the Sun were clearly important but there had to be feedbacks. Croll had first touted this back in the 19th century when he described the importance of albedo. Could there be others? Although more land with high albedo might exaggerate any cooling, the water needed to build up an ice sheet had to come from somewhere. An obvious place to look was in the oceans.

When I was a PhD student in the mid-1990s, I remember hearing a paradigm that sounded like good advice: when it got cold, the tropics were the place to be. It was felt that during ice ages, temperatures in the low latitudes didn't change much; perhaps a degree or so at most. This seemed reasonable; after all, it doesn't cool much in these parts of the world during today's winters. Perhaps surprisingly, however, this has important implications for us today. As we'll see in the next chapter, carbon dioxide levels in the air were around 30% lower during the ice age than they were immediately before industrialization. It was suggested that if the tropics hadn't cooled much when greenhouse gases levels were low, perhaps they're not as much of a problem as we might think.

Recent work has now shown this to be wrong. There's no doubt the tropics would be a great place to hang out, but they cooled by a lot more than a degree. By looking at warm- and cold-loving forams, Steven Hostetler at the US Geological Survey and colleagues have shown that tropical ocean temperatures dropped by as much as 5 °C. Other workers have shown that the high latitudes were up to 9 °C cooler than today.

These new temperature reconstructions can be used to model what effect the oceans might have had on the ice sheets. Hostetler's team took their reconstruction of ocean temperatures a step further and plugged the results into an Atmospheric General Circulation Model – or AGCM for short. We'll be looking at the results of a lot of computer models through this book so it's probably worth briefly explaining how they work. Essentially, models run a simulation of the world, using physical laws to predict the effect of changing any factor that influences climate. It's a bit like having your own virtual world. You can do what you like when you like and no one will complain. The beauty is that predictions for 100 years into the future or past don't take 100 years to compute. You can run a world with very little carbon dioxide or massive amounts and see what happens very quickly, giving insights into what the future might bring. This is done as a large 3D reconstruction of our planet with the atmosphere split up into different parcels of air. Calculations are then run to find out how much heat, moisture and wind they contain. The properties of each of these air parcels changes slightly with each calculation. The computation is then repeated, taking into account the new conditions. The model can be run for as long as you like, but it doesn't take long before you need to have a lot of computer grunt to process the numbers.

Hostetler and colleagues put their new reconstructed sea surface temperatures into the computer model to see what it would do to the atmosphere. They ran the model for the equivalent of 30 years, with the Earth's orbit set for 21,000 years ago and with 30% lower carbon dioxide levels. The results showed that the world cooled on average by over 6 °C during the last ice age. But the important thing was that the extra cooling in the tropical oceans caused less ice to melt over the Laurentide Ice Sheet, encouraging its growth. Although orbital changes may have been the driver of the ice ages, climate feedbacks clearly played an important role.

It's now clear that changes in the way the Earth orbits the Sun can put our planet into and out of an ice age. It can be quite subtle. A slight shift in the shape of the orbit, the tilt of our planet or a little wobble will change how much heat we get from our star.

The critical thing is that continents don't need to be concentrated over the tropics to plunge the world into an ice age. In themselves, the orbital changes aren't enough to cause a surge of ice over the planet, but they kick start a chain of feedbacks that greatly exaggerate the effect. These feedbacks play a critical role and it's not just in past climate change.

Although many of the ice sheets have now gone, we still have a very real hangover from this time: permafrost. Even today, permanently frozen regions cover a vast area and most of them haven't defrosted much since 21,000 years ago; in the northern hemisphere alone they make up around 24% of the surface. Recent work has looked to see what will happen to permafrost regions with future warming. The results suggest they'll retreat to the poles. Not only this, but it looks like the depth where melting takes place will increase by 30–40%. This has worrying implications on a couple of fronts. As Murton and his team found, when the permanently frozen layers melt, the greatest disruption happens at the surface; the shattered soil and rock are no longer supported by ice lenses and fall in on themselves. If you walk along the chalk downlands of England you'll often see shattered layers of rock strewn across the ground, a reminder of the last ice age. In a future warming climate, landslips may become more common as once frozen cliffs will no longer be able to support themselves; surfaces above permafrost are likely to collapse, taking with them any buildings that were constructed during more freezing times.

The changes will be terrible for those living in the permafrost but I suspect most readers of this book aren't living there. Melting permafrost doesn't obviously affect the majority of people. But there's a bigger issue here than just subsidence; the permafrost is keeping an ancient ecosystem ice-bound. In Siberia, more than one million square kilometres are covered by frozen soils known as yedoma. Over millennia, grazing mammoths have packed these sediments with carbon. Animal bones, ancient grass roots and dung have contrived to load the yedoma with 10 to 30 times the amount of carbon that's found in equivalent soils outside the permafrost. A recent calculation by Sergey Zimov at the Russian Academy of Sciences and colleagues has suggested that around 500 gigatonnes of carbon is contained in the yedoma. Once defrosted it can decompose quickly, dumping huge amounts of methane, and ultimately carbon dioxide, into the atmosphere. Worryingly, this is already starting to happen, and could be a major cause of future warming. It's a ticking time bomb.

What we need to know is whether these ancient sources of carbon are an important feedback in the climate system. Do they play a significant role in exaggerating the effects of climate change? To answer this we need to look at a time when the Earth's orbit meant that we were getting a little more heat from the Sun.

Chapter 4
A PREVIOUS WARMTH

In their 2007 Fourth Report, the Intergovernmental Panel on Climate Change summarized the results of a number of computing models that predicted the likely average temperature of the world in 2100. These models looked at several different scenarios where the concentration of greenhouse gases in the atmosphere changed over time. At one extreme, the projections included doing nothing about the problem, pumping ever more of these gases into the air – 'business as usual' – with the carbon dioxide equivalent reaching 1550 parts per million by 2100. At the other end of the scale, the low emissions scenario describes gas levels rising at first but then falling to the carbon dioxide equivalent of 600 parts per million. Up until 2030 there's not a lot to choose between them; some parts of the world will warm up more than others, while some areas may even cool; overall, the models seem to show a warming of 0.6 °C after 1990. But after 2030, the different scenarios paint widely different pictures. For the low-emissions scenario, temperatures increase on average by 1.8 °C by the end of the century. For business as usual, most of the models give a far bleaker view; they suggest that the temperature will rise on average by 4 °C.

The crucial thing we have to realize is that climate change isn't just about temperature; if it were it might be a little more manageable. One of the direct effects of a warmer planet is sea level change. Unlike warming, where the average hides a range of different trends, sea level is global and the changes will happen virtually instantaneously. Anyone living on the coast will see the effects: rich folks with their trendy beach huts and poor people living in their fishing villages. Sea level change is the ultimate leveller. The different temperature scenarios set out by the IPCC suggest that the world's sea level will rise somewhere between 20 and 60 centimetres by 2100. This would be a significant rise, but could it get any worse? To answer this we need to go back to a time when temperatures were a lot warmer than today.

In the last chapter I mentioned that during the 19th century, layers of mud were discovered sandwiched between glacial moraines. It was during the 1860s that British geologist James Geikie had stumbled upon these while tramping over hill and dale. Not only did this kick into touch the concept of just one Great Ice Age, it also showed that there had been a previous interglacial. But it was another decade before the full importance of Geikie's discovery was realized. It was in 1874 that a Dutchman called Pieter Harting looked at some cores that had been taken around the city of Amersfoort. Harting's cores were made up of ocean muds showing this part of the Netherlands had once been marine. Crucially, within the muds were seashells belonging to species that Harting was able to show did not live in the 19th century North Sea. Some of his fossil species were identical to those living in the Mediterranean, suggesting that the North Sea had been warmer than today. He named the layers 'Système Eémien' after the River Eem that flowed alongside Amersfoort. We now know these layers date back to between 130,000 and 116,000 years ago. In deep-ocean cores from around the world, isotopes in the forams show it was a period of low ice volume; a time when ice sheets were smaller than today. Could this be a vision of the future?

There's no doubt it was warm at that time. Eemian Britain, for instance, had a menagerie of hippos, elephants, rhinos and hyenas that roamed the island. It looked more like the African savannah than the rain-blurred landscape we'd recognize today. Fossil bones of these creatures have been found as far north as Yorkshire, with some of the best evidence discovered under London's Trafalgar Square. Their presence might suggest that it was blisteringly hot, but it's not as clear-cut as that. Although we might associate this wildlife with today's Africa, these beasties have often happily inhabited Britain during interglacials. The Eemian was no exception. There is no evidence for people at the same time, suggesting that our ancestors' absence from Britain allowed these exotic animals to prosper. As a result, their presence is not a great indicator of temperature. It tells us it was warm, but by how much?

A number of Eemian sites have been found across the high Arctic that shed some crucial light on what was happening at this time. Today, winter sea ice comes out through the Bering Strait and reaches as far south as the Alaska Peninsula. But on ancient sea beaches discovered around the Bering Strait, rock oysters have been found that can't tolerate icy conditions. Other fossil seashells of species found in warmer, sunnier climes have also been uncovered. It all points to winter sea ice being far less common than it is today. The Bering Strait was probably ice-free, with winter ice about 800 kilometres north of today's limits.

This is not the only evidence of warmer temperatures during the Eemian. Buried forests from this time have been unearthed in northern Alaska, rich in spruce pollen, cones and leaves. These results are exciting. By looking at fossil plant remains in ancient muds, scientists are able to reconstruct past ecosystems and use today's known distributions to infer past climate. But it's not always clear how representative changes in different pollen types are of the local environment. At an extreme, pollen can be blown into a site from hundreds of kilometres away. But fossil leaves and cones can't travel far; their presence shows that spruce was growing close to where the samples were taken. Their discovery shows that these trees were capable of growing in locations at least 80 kilometres north of their maximum extent in North America today. On the other side of the Bering Strait, in northeast Siberia, larch was also significantly further north. Some recent studies of Eemian muds show that larch was 600 kilometres polewards of today's range. Average July temperature is the main decider of where larch can thrive, suggesting summers here were 4–8 °C warmer.

But these examples are just snapshots of time that have fortuitously escaped destruction. What we really want is a long, detailed and continuous record of the climate changes that happened in the Arctic through the Eemian; something that will allow us to work out what happened when. Fortunately we have a goldmine of information sitting smack bang in the Arctic. We can turn to the Greenland Ice Sheet.

The dimensions of the Greenland Ice Sheet are staggering. Outside of Antarctica, half of all the freshwater in the world is locked up in Greenland. Of the 2.2 million square kilometres that make up this island, 85% is covered in ice. The highest point – known as Summit – is nearly 3.3 kilometres above sea level, while the sheet itself measures 2500 kilometres north to south. We're going to be looking at Greenland a lot in this book. Ice cores taken from here can be more than 3 kilometres long, stretch back 123,000 years and allow exquisitely detailed reconstructions year by year. It's probably the closest thing climate researchers have to a holy place and is almost considered the template to compare other records against.

For an ice sheet like Greenland, it's important to realize that as ever more snow is being added to the surface, the previous year's fall is driven deeper down, eventually turning to ice. At the same time that snow is accumulating, ice is melting and calving on the outer limits of the ice sheet. Because of all this, the ice flows down and out from a high ridge that runs up through the middle of Greenland, the so-called ice divide, to the edges, following a curved path (Figure 4.1). The problem with coring in locations downhill of an ice divide is that the layers can get mixed up by the flowing ice. In theory, immediately below a high point, the ice sinks straight down, preserving the horizontal layers of former snowfall with little disruption until they get close to bedrock. It's these high points on the ice surface that are the best to core because they preserve the longest and most reliable records. But given that the temperature at Summit regularly gets down to −32 °C it's not surprising that work in this part of the world took a while to get started.

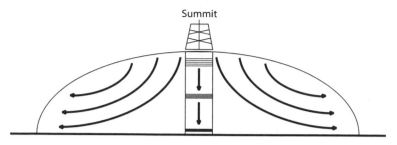

Figure 4.1 Ice flow in the Greenland Ice Sheet.

The earliest record of anyone digging in Greenland for science was Alfred Wegener, someone we met in Snowball Earth. Along with Danish officer J. P. Koch, Wegner overwintered on the ice in 1912 and hand drilled a 25-metre core, taking temperature measurements as they went. But it wasn't until 1929 that the potential was realized when Wegener led another expedition up onto the ice. His team set off explosions on the surface and because of the time it took for the echoes to bounce off the bedrock, worked out the ice where they were working must be around 2.5 kilometres thick. They also dug pits and were able to show that the snow was laid down in layers each year. Finding there wasn't enough food and drink to stay at the team's inland site, Wegener and his Greenland companion opted to return to base camp. Sadly they died during the trip. His companions on the ice didn't learn of the tragedy until the following summer.

Since the early work led by Wegner, lots of cores have been drilled across Greenland. These are incredibly well dated. In 2004, the North Greenland Ice Project – or NGRIP – reported an ice core of over 3 kilometres in length that was taken on a ridge to the north of Summit. By happy chance, the bottom of the ice is melting at the rate of about 7 millimetres per year. As the ice melts at the base of the ice sheet, the sinking layers above don't get squashed nearly as much as they usually do at these depths. As a result, a year-by-year record of ice can be counted down to a staggering 123,000 years ago. Even at a depth of 2.9 kilometres (equivalent to an age of 105,000 years), a year is represented by over one centimetre of ice. This is twice the thickness of equivalent layers in other cores of the same age and allows a huge amount of information to be extracted.

The beauty of ice cores is that they contain a direct signal of what was happening in the environment. Unlike the vagaries of some biological responses to climate, what's locked up in the ice is a direct measurement of what was happening at the time: genetic material shows what sorts of plants and animals were about, sea salt gives an idea of ice cover over nearby oceans, and certain acids can indicate when volcanic eruptions were going off around the world. If that wasn't enough, when the snow finally turns to ice, gas bubbles become trapped, preserving what was in the air at the time.

Just as the forams preserve a record of the isotopic content of the oceans, the ice preserves that of snowfall. As early as the 1950s it was realized that there was more heavy oxygen in precipitation falling in warmer climes. The reason for this is relatively simple. It's far harder to evaporate water that contains heavy oxygen. The little that gets into the air rapidly falls as rain or snow, and the consequence is that, over time, the remaining water vapour in the air contains increasing amounts of light oxygen, which eventually rains out at cooler temperatures. We now know there are complications to this relationship but a simple temperature control is a good rule of thumb: the more heavy oxygen in rain and snow, the warmer it was.

In the central and north Greenland cores it was noticed that the Eemian ice layers had a lot more heavy oxygen than snow falling in the same areas today. The size of this difference meant that if the signal was converted to temperature the Eemian was 5 °C warmer than now. Not only this, but the southern part of Greenland appears to have had little or no ice at this time. It all points to the Arctic being a lot warmer during the Eemian than in the late 20th century. This had to have been natural, so what happened?

For a lesson on what might lie in the future, the Eemian gives some fascinating insights. It's not entirely certain what the average world temperature was at this time, but it looks like it was around 1 °C warmer than today. We know that because of the Earth's changing orbit, most of this warming took place in the Arctic, a region that received a lot more heat from the Sun during spring and summer. The question was whether these orbital changes were enough to cause the Eemian temperatures.

To test this, in 2006 Bette Otto-Bliesner and colleagues at the National Center for Atmospheric Research in the USA ran a computer model to simulate the ocean, atmosphere, land and sea ice of the Eemian. They plugged into this the high amount of insolation falling over the Arctic during the Eemian and ran the scenario for the equivalent of thousands of years. The team found that significant changes fed through to many different parts of the climate system. The results were alarming.

Otto-Bliesner and colleagues found that changes in the Earth's orbit could indeed drive much of the warming. The model suggested that Siberia was 2–4 °C warmer than today, with a warming of nearly 3 °C across central Greenland. More importantly, the results showed there was a lot less sea ice. During the peak in warming, the amount of summer sea ice in the Arctic was 50% less than today; in the Eemian, most of it was north of 80° latitude.

To build the vast ice sheets we know existed during the ice ages, the water had to come from the ocean. As a result, sea levels crashed; the latest estimates indicate that it was somewhere around 120 metres below where it is today. The result was that when an ice age was in full swing, vast areas of the continental shelf were exposed, joining up islands and continents around the world. Looking from above, towering ice sheets would have dominated the high latitudes of the northern hemisphere while the continents would have become larger than today. When the

Eemian occurred, the Laurentide and Scandinavian ice sheets melted, catastrophically flooding the exposed continental shelves and raising the sea to levels we'd be more familiar with today.

Because there was a lot less sea ice in the Arctic, more of the ocean was opened up to heating, with the result that the Greenland Ice Sheet and other smaller ice caps in the Arctic would have started to melt. The team modelled what this would have done to the sea level. It looked like less ice in Greenland would have pushed the tide higher by 3.4 metres.

What sort of processes can melt this much Greenland ice? The critical thing appears to be the timing of the warming in the seasons. If melting starts to take place in the spring, the surface snow becomes slush earlier and lasts for a greater proportion of the year. As we saw with Snowball Earth, wet snow has a much smaller albedo than the fresh stuff. In fact, only around 40% of the Sun's radiation is reflected back off the surface. This can have a huge effect on the melting of ice. The high insolation that we know happened during the Eemian spring and summer would have greatly reduced the albedo of the surface, allowing much more of the Sun's energy to be absorbed. As the surface melted, the overall height of the ice sheet would have dropped, exposing it to the warmer temperatures of lower altitudes. And just as small cuts of meat take less time to thaw than the equivalent weight in one joint, more of the ice sheet's volume would have been exposed to warming, speeding up the melting.

But heating of snow and ice can also play another trick that was first explained by one of the great and most entertaining characters in science: Benjamin Thompson (or, to go by his later title, Count Rumford). Although born in Massachusetts in 1753, he was an ardent monarchist and became a British army officer during the American War of Independence. When everything had gone belly up, he fled to Britain, got himself promoted to colonel and was later instrumental in setting up the Royal Institution in London. A great scientist and businessman, he was also a successful leader; often all at the same time. In spite of his

American–British background, he became a serving officer in the army of Carl Theodore, Elector of Bavaria, and stayed in Munich for 12 years. In 1796, converging French and Austrian armies threatened Munich, and the Elector, fearing for his city, begged Rumford to return from London. Before he knew what had hit him, Rumford was put in charge of the defence of a city that faced two large opposing armies camped outside. Most of the great and the good – including Carl Theodore – had fled out of the war zone. Rumford rallied the defending forces, built up the defences and played for time, negotiating with the French and Austrian armies. Against expectations, the ploy worked: the French and Austrians retired and Rumford suddenly found himself the toast of Munich. He certainly led a charmed life.

In the early 19th century, Rumford took a tour of central Europe and made a fascinating observation. In the summer months, locals had observed narrow, deep pits of water on the surface of the Mer de Glace – the Sea of Ice – a glacier on the north side of Mont Blanc. At the time it was unknown what caused the pits to form each year. In 1804, Rumford reported to the Royal Society that he had made some new insights into this intriguing phenomenon. Generally speaking, liquid water becomes denser at lower temperatures. But it's not at 0 °C that water reaches its maximum density; Rumford had been one of the first to realize that it happens at 4 °C. The practical upshot of this is that if ice melts, the density of the water increases up to a temperature of 4 °C. Rumford realized the pits on the surface of the ice must be due to the effect of changing temperature on the density of water. He argued that small pools of water would form on the surface of the ice at the beginning of summer. With a temperature just above freezing, the surface pool water would have a relatively high density. This would then sink to the bottom, melting any ice around the edge. Over the summer, the pool would deepen as dense, 'warm' water continued to form on the surface and then sink to the bottom. The more warming, the more pits would develop and the bigger they would become. A similar process would have helped the Arctic ice to melt rapidly during the Eemian.

These sorts of changes wouldn't have been enough to explain all of the sea level rise known during the Eemian. Otto-Bliesner's team showed that melting Arctic ice could explain at most a rise of 3.4 metres. But it's known that sea levels were 4 to 6 metres higher than today. There are various ways to reconstruct what sea level was doing in the past. For example, ancient beaches give an upper limit to tide levels, while many coral species only survive in shallow water. But because of plate tectonics, you can't just look at any old coral or beach. You have to home in on areas of the world that haven't moved up or down over time or probe areas where the amount of movement is known. By careful selection, fossil corals and beaches dating back to the Eemian point to the world's sea level being 4 to 6 metres higher than they are now.

Otto-Bliesner and colleagues knew that the melting of the Greenland ice couldn't explain all of the higher sea level. In a companion paper headed by Jonathan Overpeck from the University of Arizona, they looked to see where the rest of the sea level rise came from. They found that most of the warming was limited to over the Arctic. Elsewhere, temperatures were only a little higher. Because water expands when it warms, this result had an important implication. It meant that the rest of the sea level rise couldn't have come from a warming ocean. This only left one place. The rising seas had to have come from melting Antarctic ice.

Overpeck and colleagues calculated there was little extra summer insolation over Antarctica during the Eemian. So why did the ice collapse? Well, it's not just by surface heating that the ice can melt. The Antarctic Ice Sheet can be split into two quite distinct parts: west and east. The Eastern Antarctic Ice Sheet sits on ground above sea level. The Western Antarctic Ice Sheet, by contrast, is relatively smaller and lies to the west of the Transantarctic Mountains; it reaches out into the sea where it is buttressed back by vast ice shelves. Because most of the land in western Antarctica is so low, much of the inland ice sits on ground that is below sea level. These ice shelves make the Western Antarctic Ice Sheet particularly vulnerable to warming and changes in the ocean.

A sea level rise of up to 3.4 metres would have conspired to lift the West Antarctic Ice Sheet off the ground. This would have allowed seawater to get in under the ice streams. Overpeck and colleagues modelled a warming in the Antarctic waters of between 0.5 and 1 °C during the Eemian. Melting would have taken place underneath the ice sheet, thinning the overlying ice. Once the ice shelf had started to collapse, the rest of the ice sheet would no longer have been buttressed back; glacier streams would have accelerated towards the coast. Even just a small increase in temperatures could have been enough to cause a devastating change. Today we know that once this process starts to happen, more crevasses open up on the surface of ice, allowing surface pools of water to drain below, lubricating the bottom of ice streams. It's just one feedback built on another. Overall, there appears to have been enough going on to drive melting and calving of the Antarctic ice to make up the extra sea level rise to 4 to 6 metres.

It was more than the poles that were warm during the Eemian. In the oceans, the forams have yet another surprise for us. Although Emiliani had been wrong that the oxygen content of the foram shells was driven solely by temperature, another component is. Instead of calcium, forams can also take up magnesium to help form the carbonate in their shells. This is temperature-sensitive: the more magnesium in the shells, the warmer it was. The beauty is that it's accurate to within a degree. There are some potential problems, but basically a consistent pattern emerges with the surface-dwelling forams: during the Eemian the tropics appear to have been 1–2 °C warmer than today. Along the west Australian coast, Eemian corals have been found up to 500 kilometres beyond their current range – some as far south as the coastal town of Esperence. This all suggests that the Leeuwin Current that bathes the Western Australian coastline must have been warm enough to bring coral larvae all the way down from northern Australia and/or Indonesia. If orbital changes weren't helping to make these latitudes much warmer, might greenhouse gases?

To answer this we need to go the ice cores. In November 2006, I visited Jørgen Peder Steffensen at the Niels Bohr Institute in Copenhagen. Better known as JP to his friends, he took me inside a huge walk-in freezer. Archived in the frigid –20 °C air were cores of ice collected from across Greenland. I remember being excited when JP pulled out a core of ice measuring just 55 cm long. It was labelled number 910 and had been taken 500 metres below the Greenland surface. In it was ice that was laid down between 1 BC and AD 1. Peering closely I could see small gas bubbles entombed in the ancient snowfall. These bubbles should contain air from over two millennia ago. Or do they?

As a glacier or ice sheet builds up over time, the snow that fell on the surface is gradually buried. As it does so, the flakes gradually get

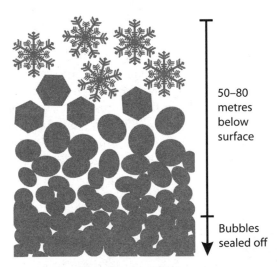

50–80 metres below surface

Bubbles sealed off

Figure 4.2 The firn.

squeezed ever more closely together. When this old snow – or firn – eventually gets down to between 50 and 80 metres, enough weight is overhead that it forms ice (Figure 4.2). Before this happens, there are enough gaps between the old flakes that the air within the upper 50 metres or so can still escape to the surface and the air from the surface can get down into the gaps between the old snow flakes. It's only when the old surface finally gets down below 80 metres that the bubbles finally become pinched shut and trapped. Escape is no longer an option. The gas bubbles are entombed and contain a snapshot of what was in the air. The problem is that the gas is not the same age as the ice that surrounds it.

Fortunately this offset can be corrected for. If you can date the ice precisely, the depth of the firn – and therefore the age difference – can be determined. Because of the relatively high levels of snowfall, the Greenland firn becomes closed off within 300 years. Unfortunately for Greenland, carbonate dust from the deserts of eastern Asia is blown onto the ice, screwing up the carbon dioxide in the bubbles; the other main greenhouse gases – such as methane and nitrous oxide – remain unscathed. In Antarctica, no significant amounts of carbonate dust are blown onto the ice and because the atmosphere is so well mixed, past changes in the amount of atmospheric carbon dioxide can be reconstructed. So if we want to get a good handle on what carbon dioxide was doing in the past we need to go to Antarctica. Just as in Greenland, a multitude of cores have been taken across the ice sheet. The longest record has been taken at the eastern Antarctic site known as Dome C, where a core measuring 3.2 kilometres has just been drilled by the European Project for Ice Coring in Antarctica (fortunately this mouthful is often abbreviated to EPICA). Because snowfall is so much lower over Antarctica, the record spans over 800,000 years, reaching much further back in time than Greenland. Future drilling projects hold out the promise that ice will be recovered that is even older, possibly as much as 1.2 million years old.

Before the Dome C record, the longest Antarctic record was taken from the coldest place on the surface of our planet; an ancient Soviet drilling site known as Vostok where the temperature averages –52 °C. Because of its location and sub-freezing temperatures, Vostok sits on top of ice that stretches back hundreds of thousands of years. In the late 1980s, the researchers had got back to 160,000 years ago. By 1999, the team had got back 420,000 years.

In these cores was a continuous record of change. In the process, researchers found ancient microbes that had been deep-frozen in the ice for several hundred thousand years; against all the odds these were revived. Vostok also had other surprises: immediately below the coring site was discovered a vast freshwater lake that has become known as Lake Vostok. This is the largest known lake under the Antarctic ice, measuring more than 250 kilometres long; more than 100 have since been discovered. Russian scientists are currently preparing to explore these lakes, which hold the promise of life that hasn't seen the light of day for hundreds of thousands of years. Who knows? It may lay the groundwork to discover life under the Martian polar caps.

More relevant to our story are the isotopes in the Eemian ice which show that it was also warmer in Antarctica than today. Because carbonate dust isn't the problem that it is in Greenland, the cores also contain a beautiful record of changing carbon dioxide levels over time. It's clear that the levels of carbon dioxide in the air have kept within a narrow range. During an ice age, the amount of gas got down to around 180 parts per million. During interglacials – before humans had a big impact on the atmosphere – the levels seemed to have been between 280 and 300 parts per million, showing that when it got warm there was more greenhouse gas in the air. But there's an important point to make here. The carbon dioxide – and the other greenhouse gases for that matter – did not cause the first rise in interglacial temperatures. Once the correction has been made for the firn depth, there seems to have been a time lag of several hundred years. Gas levels didn't start increasing in the atmosphere with the first warming. The implication is that the greenhouse gases were not the primary driver of the big climate cycles but are an important feedback. This fits in nicely with what we've learnt about orbital changes.

The big question for us in a greenhouse gas-rich world is how much of a feedback does carbon dioxide have? To get a handle on this, it's useful to look at what temperatures would do with a doubling of atmospheric carbon dioxide relative to pre-industrial levels; something described as the climate sensitivity. The traditional view is that when the land, oceans, ice and atmosphere have reached a new balance with a doubling of carbon dioxide, the average temperature of our planet will be anywhere between 1.5 and 4.5 °C warmer. To test this, Margaret Torn at the Lawrence Berkeley National Laboratory and John Harte at the University of California investigated the record from Vostok. The results were an eye opener.

Torn and Harte looked at the levels of carbon dioxide and methane trapped in the air bubbles spanning the past 400,000 years within the Vostok ice. From this they could compare changing temperature with greenhouse gas concentrations and were able to show that there was an extra warming effect. The greenhouse gases were amplifying the first Eemian warming caused by the extra heat from the Sun. Worryingly, however, extra warming seemed to increase greenhouse gas levels further. When they combined these results with those produced by climate models they were able to propose a new climate sensitivity: it was somewhere between 1.6 °C and 6 °C. This means that for a doubling of carbon dioxide to 550 parts per million, global temperatures could increase by as much as 6 °C, frighteningly higher than most climate models are predicting at the moment (though some models have suggested it could be as high as 11 °C). It implies that the climate system may be far more sensitive to carbon dioxide than is currently accepted.

So what could cause this extra feedback between temperature and carbon dioxide? It's important to realize that when greenhouse gases enter the atmosphere they don't just sit there; they're taking part in a bigger process. A complex series of exchanges take place between different carbon reservoirs in the ocean, land and sky that make up the carbon cycle. These flows of carbon work to a number of different time-scales with the result that most of the carbon dioxide entering the atmosphere disappears within a century, although some 20% may hang around for millennia. We saw earlier in Snowball Earth that smashed rocks can act as a sink, stripping the air of carbon dioxide as it works to break them down. This is a terribly slow process, taking tens of thousands of years, and doesn't look like it had much impact on greenhouse gas levels during interglacials and ice ages. But the oceans and land can also work on a faster time-scale, rapidly changing the amount of carbon dioxide in the air. Surface ocean waters, for instance, directly absorb the gas from the atmosphere. Living organisms can also fix carbon dioxide via photosynthesis. In the oceans some of this carbon is recycled by other organisms eating the algae's dead remains, but a significant portion sinks, where it either gets locked up in the sediments on the seabed or is released in the deep as carbon dioxide. Locking up the carbon in the ocean depths in this way is known as the biological pump, and it may have been a reason why carbon dioxide levels were so low during ice ages; dust levels in the atmosphere are estimated to have been 2 to 20 times greater at these times and may have provided the essential nutrients for the algae, acting as a fertilizer that allowed them to thrive in the surface waters and suck carbon dioxide out of the air.

So to get a carbon feedback of the likes that Torn and Harte suggest, these flows of carbon had to have been interfered with or even reversed. When water warms up, its properties change. Not only does it expand, but its capacity to hold onto dissolved gases becomes less. The principle is just the same as leaving a bottle of fizzy drink out in the midday Sun; the dissolved carbon dioxide that gives the drink its fizz becomes desperate to escape at higher temperatures. The result is that a turn of the cap allows the gas to escape. Seawater may not be so explosive but it has about fifty times more carbon dissolved in it than air. With increased warming, the ocean may also circulate more intensively, bringing dissolved carbon dioxide back up from the depths and pumping it into the air. The Southern Ocean that bathes Antarctica looks like it's a particularly important player. The warmer it becomes, the more carbon dioxide comes out of the deep sea and goes into the atmosphere. But the action isn't just happening in the oceans. On land, plant matter – including carbon sequestered in the soil – breaks down more easily with warmer temperatures, acting as another significant positive feedback.

The key thing here is that just because carbon dioxide lags behind temperature on the time-scale of interglacials, it doesn't mean this gas can't trigger future climate change. The results by Torn and Harte show that greenhouse gas levels greatly amplified the temperature changes started by orbital variations. Carbon dioxide was clearly an important player in the climate system of the recent geological past.

It's clear that even though orbital changes were the initiator of long-term climate changes, feedbacks played a crucial role in exaggerating their effect. Albedo, sea level and greenhouse gases all helped make the world much warmer during the Eemian. This has important lessons for the future. The European Union has set a target that the world's average temperature must not warm by more than 2 °C. This might not sound much but remember that it's an average. When you consider the regional extremes in temperature and the jump in sea level that took place during the Eemian when the world was only around 1 °C warmer, the importance of the European Union target is all too clear.

It is possible that life itself may have a more direct influence on climate than it's generally given credit for. British scientist James Lovelock has made a convincing case that living organisms regulate the world's climate so as to keep the planet inhabitable. This is the exciting Gaia Hypothesis. The idea is that living things are capable of changing a whole host of different aspects of our planet, including greenhouse gas levels, cloud cover and albedo, thereby making the Earth a self-regulating entity. Recent research suggests that many of these processes do indeed take place. I won't go into this in detail because there are a host of excellent books on the topic. Lovelock's recent book *The Revenge of Gaia* is well worth a read, although his predictions for the future are not optimistic. In it he argues that Gaia has been pushed too far. Life cannot correct for the changes we've started and we're on the verge of a series of catastrophic positive feedbacks that will greatly magnify the warming we've initiated. For our sakes I hope he's wrong.

At the moment, the ocean is doing a good job of soaking up our emissions; it's estimated that the world's seas have been taking up around half of all the carbon we're dumping in the atmosphere. The net effect is that there's less carbon dioxide in the air than there should be. Up till now the Southern Ocean has been a particularly good sink for carbon dioxide, soaking up the gas and pumping it deep underwater. Worryingly, it now looks like this system might be becoming less helpful; recent work in the Southern Ocean is showing that between 5 and 30% less carbon dioxide is being absorbed today. It appears to be reaching a saturation point. Other studies indicate that gas levels in the air are starting to increase faster as the Earth's natural sinks for carbon dioxide become less effective. Climate models are now taking carbon cycle feedbacks into account and the results are disconcerting. By the end of the computer model runs, the oceans are indeed taking up less carbon and the soil is releasing more; just as the ice core work by Torn and Harte implies. The bottom line seems to be that high greenhouse levels in the atmosphere cause warming which further increases the amount of carbon dioxide in the air. The climate models suggest that global temperatures can increase by up to another 1.5 °C with this feedback.

If these sorts of temperature change do take place in the future, just how fast will the ice melt? Today's studies make sobering reading. Over the past few decades, temperatures across the Arctic have been rising on average by about 0.5 °C every ten years. This has been accompanied by a decrease in the extent of summer sea ice, with a loss of more than 7% per decade. It's not quite clear why the Arctic is so vulnerable to warming, but there are probably a multitude of reasons. Things aren't helped by the fact that the atmosphere over the Arctic is about half the thickness of the tropics; as a result, it doesn't take as much energy to heat up the air. If this wasn't enough, the continued release of pollutants into the air – including soot – is severely exacerbating the situation. Pollutants can act as a double whammy: not only do particles reduce the albedo of the snow, they can also directly absorb any heat that's about.

This all has an effect on ice sheets and sea level. But modelling these changes isn't easy. For a start, although we know that higher temperatures will cause the seawater to expand, by how much is still unclear. Moreover, at the time the Intergovernmental Panel on Climate Change were putting their report together it was uncertain whether the rate at which ice is being lost from Greenland and Antarctica will change in the future; this was largely because studies of past ice flow rates were few and far between. As a result, the models used by the Intergovernmental Panel on Climate Change had to assume that ice loss would continue to be the same as today (though with a big warning light over this assumption). The result was the predicted sea level rise of 20–60 centimetres by the end of the century. Unfortunately, recent analysis of satellite data shows there are increasing losses from Greenland and Antarctica, largely driven by more ice melting through the summer.

Fortunately, where the models become less certain, the past steps forward. To try to get around the uncertainty of future sea level rises, Stefan Rahmstorf at the Potsdam Institute for Climate Impact Research has used known changes in temperature and sea level over the past 3 million years to calculate the general relationship. This suggests a change in sea level of 10–30 centimetres per degree. In the short term, expansion of seawater will probably dominate the rise in water level. By Rahmstorf's reckoning, a sea level rise of between 50 and 140 centimetres is likely by 2100. This might not sound much, but a sea level rise of 1 metre will directly affect at least 145 million people. Throw in an increasing number of storm surges and the likely changes in coastal erosion and the number of people affected will increase significantly. As temperatures continue to rise, the fear is that the world's ice sheets will play a larger role. The concern is a big one. In 1978, John Mercer from the Ohio State University caused a minor panic when he suggested that a doubling of carbon dioxide could cause the West Antarctic Ice Sheet to disappear, raising sea levels by 5 metres. It was unclear at the time how long Mercer thought this might take to happen, but it was implied to be quick.

Modelling work by Overpeck and colleagues suggests that Mercer wasn't totally out of left field. By their calculations, once the ice sheets start melting in earnest, a rate of rise of more than 1 metre per century is not unlikely. Recently, James Hansen at the NASA Institute for Space Studies and colleagues have taken this a step further and suggested that the sheer size of the ice sheets won't be enough to hold back the effects of warming forever. If the ice sheets continue to melt for a century it could result in their collapse. Modelling of future climate suggests that temperatures may be high enough by the end of this century to irreversibly melt the Greenland ice over the next millennia. High temperatures mean high sea levels and they're looking more likely by the day. We'd better get ready to batten down the hatches. Is there anything out there that might help cool the planet down?

Chapter 5
ATLANTIC ARMADAS

Up till now we've largely looked at what drives long-term climate change. But there is a worrying gap in our knowledge. Most of what we've covered has been gradual; we've not really seen any abrupt changes. It's probable that the warming 55 million years ago happened very fast but we can't date the ocean records accurately enough to demonstrate this. Are abrupt shifts in the climate system possible? Can it be green and pleasant one year and hot and dry the next?

To answer this we need to come forward in time and look at the North Atlantic after the Eemian. Originally it was thought that when things cooled down around 116,000 years ago, the ice sheets of the world expanded over tens of thousands of years, eventually reaching their largest size during the Last Glacial Maximum 21,000 years ago. We now know it's a little more complicated that this. The ice sheets didn't simply grow out, consigning more and more of the land to an icy domain. Instead, they took a few steps forward, followed by a few steps back. There were times of cooling and there were other times of dramatic warming.

The first real hint that this might have happened was only reported two decades ago. In 1988, German oceanographer Hartmut Heinrich published a research paper looking at some cores of ocean mud taken from a deep-sea hill on the seabed in the eastern North Atlantic. The cores were packed with forams but now and again layers of rubble interrupted the muddy sediments. Something big appeared to have happened in the region.

The idea of rubble at the bottom of the North Atlantic might seem a little odd. The cores were taken too far from any rivers that might have delivered the large mineral grains Heinrich had found. The only other possibility was that these particles had come from the ocean surface. Heinrich thought these layers might have formed from melting ice at the surface. But he wasn't suggesting something akin to a Snowball Earth. As we saw earlier, glacial ice can contain an eclectic mixture of things that have been plucked from the landscape. Once a section of ice has worked its way through a glacier or ice sheet, it may reach the sea where it calves off to form an iceberg. Heinrich envisaged armadas of ice being cast adrift across the North Atlantic. As the icebergs followed the ocean's currents, they'd gradually melt and any rubble contained therein would rain down to the sea floor, forming a layer rich in ice-rafted debris (Figure 5.1).

We now know that there's always the odd iceberg floating around in the North Atlantic. But one or two blocks of ice couldn't explain the amount of rubble found by Heinrich. Instead, he argued that at several times in the past, massive releases of icebergs happened across the North Atlantic. As these fanned out

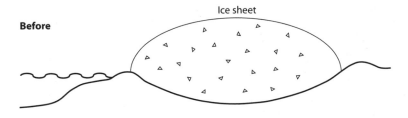

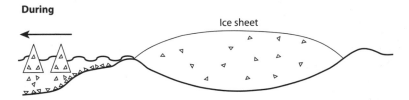

Figure 5.1 Before and during a Heinrich event.

across the ocean they melted, blanketing the sea floor with rubble. When these layers were analyzed in detail, it was realized that very few contained forams. This suggested that the pulse of rubble had fallen to the seabed very quickly, diluting the number of forams that were going the same way. But just when it looked like they were here to stay, the rubble layers ended abruptly, to be covered by more typical ocean muds and forams.

The layers seemed to form every 11,000 years or so. Most of the time not a lot happened, then suddenly a salvo of rubble would rain down from above. After 11,000 years another layer formed. Heinrich reported ten distinct layers in the ocean muds since the end of the Eemian. He thought there might be some link to the way the Earth orbits the Sun. But if so, it was clearly having an usual effect. Nothing seemed to happen and then bang: huge amounts of rubble fell into the ocean. It was all a bit odd.

It's often the case in science that when a revolutionary paper is published, its importance is not at first appreciated. In this case it wasn't quite as bad as tumbleweed blowing down the street, but not far off. Not many people seemed to know what to do with the results that Heinrich reported. The idea just hung there. It was only in 1992 that a series of papers reported the same layers of rubble in other parts of the Atlantic sea floor; they were called Heinrich events after their discoverer. The rest of the community sat upright and began to pay attention. The floodgates opened. Researchers suddenly started finding these layers all the way across the North Atlantic, some as far south as the Portuguese coast.

It soon became apparent that the layers of rubble were thickest in the northwest Atlantic; around the Labrador Sea, layers several metres thick have been found, while across in the eastern Atlantic they're just a couple of centimetres thick. In almost all the layers, limestone was a common mineral, but other types of grains have also been found which date to 900 million years old. This isn't to say the armadas of ice happened nearly a billion years ago; instead it shows that the ice had picked up rubble that dates back to before the Cambrian. All the evidence pointed to the ice coming off the Canadian part of the Laurentide Ice Sheet, probably from the Hudson Bay region. But what could have caused these big splurges of ice?

Back in the 1970s, Willi Dansgaard, Hans Oeschger and Chester Langway had realized the Greenland ice cores contained a different record of climate change to that obtained from the sea. Reconstructions of climate from the oceans seemed to show a gradual cooling trend after the Eemian, culminating in the Last Glacial Maximum. Instead, the ice cores seemed to show massive swings in temperature over Greenland; increases of up to 16 °C were recorded within a few decades. The temperature changes had a stepped pattern. Rapid warming was followed by gradual cooling, which then ended with another bout of rapid warming. Twenty-five of these warm events have now been identified and were called Dansgaard–Oeschger events in honour of two of their original discoverers. These wildly swinging changes in temperature happen too quickly and were too short in duration to have been driven by the changing orbit of the Earth. As a result, the terms *interglacial* and *glacial* are inappropriate and instead these shorter episodes of warming and cooling are known as *interstadials* and *stadials*. It was unclear how they could have occurred, especially since the ocean didn't seem to be dancing to the same tune.

In 1993, Gerard Bond and colleagues at the Lamont-Doherty Earth Observatory at Columbia University undertook some new work to try to make sense of the different patterns seen across the North Atlantic. They reasoned that the oceans needed to be reinvestigated. When developing climate reconstructions, there's often a trade-off between the length of record and how finely it can be sampled. It's unusual in nature to have very long records with year-by-year detail. Typically it's a case of researching an area and taking what you can get. In the ocean, many sequences of muds accumulate quite slowly; its often only possible to build up a coarse picture of long-term changes in the ocean because the detail isn't there. This can be a major problem. Imagine listening to a chat show on radio: the whole broadcast allows you to understand a point of view being put across. But if a filter is added that only lets you hear every second or third word you'd struggle to get the full story; you might get the gist but the detail would be lost. The same is true of climate records. If you are only getting a bit of the story every thousand years or so, it's hard to work out what's happening in between.

Bond's team focused on several records across the North Atlantic sea floor that could be sampled every few hundred years. The result was they had a far better idea of what was happening in the oceans over time. Looking at the proportion of cold- and warm-loving forams in the ocean muds they were able to recognize big shifts in climate. Sometimes there were lots of cold-loving forams; during other periods, the population of these types crashed and the forams were dominated by warm-loving species. Bond's team showed that the peaks and troughs in different forams agreed with the warm and cold temperatures swings in Greenland. The interstadials and stadials had indeed happened on land and in the sea; their absence in the ocean had just been a sampling artefact. The filter had been set too high on the earlier cores. The oceans were now showing that temperatures had dropped by up to 10 °C during a Heinrich event.

But Bond and colleagues took this work a whole step further. Not only did they have Dansgaard–Oeschger events inside their ocean cores, but they also found the same layers of rubble that Heinrich had described back in 1988. Here was a chance to put the jigsaw puzzle together. The warm interstadials and the cold stadials could be used to link up Greenland and the North Atlantic records. It was a classic bit of work. The records showed there were cycles working within cycles: a rapid warming would be followed by a gradual cooling over several hundred years, then followed by another abrupt phase of warming (Figure 5.2). The important thing was that successive interstadials didn't get as warm as previous ones; each time a warming happened, the amplitude got smaller. After several of these ever-decreasing

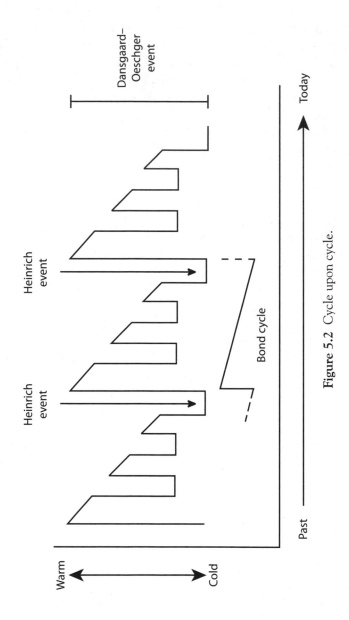

Figure 5.2 Cycle upon cycle.

cycles, the temperatures got to a low point and then culminated in a Heinrich event: the lowest temperatures seemed to coincide with a massive splurge of ice across the Atlantic. A vast armada of ice was released, creating a layer of rubble on the ocean floor. After several hundred years, the ice was spent and the climate system reset itself: the temperature suddenly increased to a maximum and the whole thing repeated. This saw-tooth pattern of large warming followed by gradual cooling and culminating in a Heinrich event became known as a Bond cycle.

Importantly, since this early work, the Heinrich layers have also been found to contain rubble that has a different geology to the Hudson Bay; it looks like some icebergs originated from other regions around the North Atlantic. This means that it wasn't just the Laurentide Ice Sheet that was letting loose with ice; other ice sheets over Greenland, Iceland, the British Isles and Scandinavia were up to a similar trick. This wouldn't have been the odd berg floating on the surface looking to sink a *Titanic*. It would have been the equivalent of around one million cubic kilometres of freshwater. It must have been an awesome sight.

In some ways science is a little like fashion. Favourites come and go. Heinrich events were no exception. They suddenly became the next big thing to work on. A crucial first question was what had caused these massive changes? In truth, it's still not entirely clear. One of the first popular ideas to be proposed was the binge-and-purge hypothesis suggested in 1993 by Douglas MacAyeal at the University of Chicago. The idea was that, over time, the Laurentide ice became so large it could no longer sustain itself and would catastrophically collapse. Once the ice sheet had reached its maximum size, lots of sediment would have built up at the bottom; MacAyeal suggested that these sediments became slippery with water until they could hold back the ice no more.

The result: a vast amount of ice slid into the Hudson Strait. The ice built up – the binge – and then collapsed – the purge. An analogy is having a curry with lots of beer on a Friday night. You get carried away and binge, eating far too much. Several hours later you have to purge your body of the offending excess. In a similar way, binge-and-purge argued that the ice over Hudson Bay took several thousand years before it would become big enough to collapse. Implicit in this idea is that the Laurentide Ice Sheet worked to its own time-scale, regardless of what was happening anywhere else. Although binge-and-purge sounds like a pretty neat explanation, it unfortunately doesn't explain everything.

In 1995, Gerard Bond and Rusty Lotti, both at Lamont-Doherty Earth Observatory, looked at several ocean cores in greater detail. They painstakingly analyzed the cores for forams and rubble. Not only did they count the amount of rubble through each core, but they also identified the different types of minerals. This can give a good indication of where the icebergs had come from; different mineral grains are characteristic of the different land masses around the North Atlantic. Not only were there bits of Canadian limestone, but there was also volcanic glass – mostly originating from Iceland – and magnetic mineral grains – mostly from southeastern Canada around the Gulf of St Lawrence.

Dating individual layers across the North Atlantic suggested that most of the ice sheets were responding at the same time. Bond and Lotti were able to test this. By looking at the proportions of the different mineral grains within closely sampled levels, it was possible to sort out the order of events. In spite of first appearances, it didn't seem that the ice surrounding the North Atlantic was collapsing at the same time. There were subtle differences in timing. Bond and Lotti found the characteristic peaks of limestone rubble from the Heinrich events that came from the Hudson Bay region. The crucial thing was that they also found distinctive peaks between the Heinrich events. These seemed to relate to the stadials found in the

Greenland ice cores. But unlike the massive Heinrich events, these smaller rubble levels had little, if any, limestone. Instead they were filled with volcanic glass and magnetic rocks. This all pointed to the release of icebergs from other regions around the North Atlantic, most probably Iceland and the Gulf of St Lawrence.

As each short cold snap happened, it seems that bergs from the smaller ice centres around the North Atlantic poured into the ocean. When it warmed up they stopped. At the beginning of each Heinrich event, these smaller discharges of ice started up again. But unlike previous smaller cold snaps, they were followed by a massive collapse of the Laurentide Ice Sheet over the Hudson Bay. This dumped vast amounts of limestone and other minerals onto the ocean floor, swamping the smaller amounts of mineral grains that had been deposited by icebergs from elsewhere. If the iceberg release from around the North Atlantic led to the collapse of the Laurentide Ice Sheet, the binge-and-purge hypothesis couldn't be the full story.

These results had some intriguing implications. The timing of these events is not easy to date in the oceans, but if we assume the cold stadials in the Greenland ice occurred at the same time as the relatively small flotillas of bergs, then the latter seem to have happened every 1500 years or so. After several small collapses had taken place, a massive Heinrich event then took place. This all suggested that the ice over the Hudson Bay was dancing to a much slower rhythm. It had to get big enough before it could become unstable; once it was large enough, the purge happened. The critical thing was to work out what started the purge. The fact that the other ice sheets seemed to consistently lead the Hudson Bay icebergs suggests that the Laurentide Ice Sheet didn't collapse on its own accord. Bingeing doesn't look like it was enough in itself. Something else must have driven the big changes.

One possibility is a change in temperature. If it got warm enough, perhaps the ice could have catastrophically melted, ensuring all hell broke loose. The problem with this idea is that the warming in the North Atlantic only seems to have taken place after the icebergs had been and gone. Intriguingly, it's not even clear that the armadas of bergs were the initial cause of cooling. Some of the ocean records from this time show that it was getting cold before the rubble started falling to the seabed. It all seems counter-intuitive. At a time when we hear that warming might cause catastrophic melting of ice sheets in the future, the opposite seems to have been the case during Heinrich events.

As we saw in the last chapter, big collapses in ice sheets can have a large effect on the world's sea level. If huge amounts of icebergs were released into the North Atlantic, it's not out of left field to suppose that there was a change in the sea level around the world. Could a rise in sea level have caused the Laurentide Ice Sheet to fail, releasing yet more water into the ocean and driving the tide higher?

Coral reefs are often referred to as the dipsticks of the ocean. They're one of the best things for working out what sea level was doing in the past. Tropical corals rely on a rich mixture of heat, light and nutrients. As a result, they largely grow within relatively shallow water. Later on we'll look at what climate information we can get from corals, but for the moment let's just focus on what it means when we find an ancient reef.

Probably one of the best known places for ancient coral reefs is the Huon Peninsula on the eastern coast of Papua New Guinea. These fringing reefs have been worked on since the 1960s, most prominently by John Chappell at the Australian National University in Canberra. The Huon Peninsula is in one of those rather special parts of the world where the conditions are ideal for tropical coral reefs to grow, but at the same time is on the edge of a plate boundary. Stretching back at least 250,000 years, the Huon Peninsula has been rising to the tune of 3 metres per thousand years. The result is that any corals that grow on the submerged shoreline of the Peninsula eventually get thrust up into the air where they're saved for a scientist to come fossicking about. Seen from above, the Peninsula look like an enormous set of vegetated steps; the fossil coral terraces appear to stretch away forever, preserving ancient lagoons and barriers that formed during rises in sea level.

John has undertaken a vast amount of work on the Huon Peninsula. Over the years, he has led a team that has measured, prodded and analyzed most of this ancient seascape. In 2002, he reported a study which showed that when a correction had been made for the rising land, the large reef terraces grew in sea levels that jumped 10 to 15 metres during Heinrich events; the deepening water gave the corals head-space to grow at pace with the rate of rise. After the sea level had shot up, the Huon Peninsula continued its inexorable rise, preserving a record of the big changes that had taken place. As the ice sheets subsequently reformed, the water was sucked back out of the ocean and the tide fell. Intriguingly, no sea level jumps were detected at the Huon Peninsula for the smaller stadials that happened between the Heinrich events. The collapse of the smaller ice sheets might have caused sea level changes of less than 3 metres; probably beyond what the coral reefs on the Huon Peninsula can detect. Worryingly, more recent reconstructions imply these estimates may be conservative. Work in the Red Sea suggests that sea level changes of up to 35 metres might have taken place during the Heinrich events. These are gargantuan changes.

It's still not agreed how much ice was lost to the sea during Heinrich events and the other stadials, largely because there's little evidence preserved on land. Modelling studies suggest that the height of the Laurentide Ice Sheet dropped by over a kilometre during a Heinrich event and lost around 15% of its volume to the sea. It's not yet clear whether these numbers are correct or that they could explain a sea level rise of up to 35 metres. If we accept that such a large sea level change did indeed happen during Heinrich events, an obvious extra source of water is the Antarctic Ice Sheet.

In 2000, Sharon Kanfoush and colleagues at the University of Florida looked at whether there was ice-rafted debris in South Atlantic ocean muds off Antarctica. In spite of looking at cores over 12 degrees of latitude, there were several distinct layers of rubble that seemed to be common to all the records. This showed that the Antarctic Ice Sheet was not one unresponsive lump that remained unchanged over time. The vast ice sheet in the south did indeed wax and wane. It seemed to have experienced similar changes to those tumultuous events in the North Atlantic. The big problem was dating the South Atlantic layers so they could be compared to the climate changes preserved in the Antarctic and Greenland ice cores. Were these layers being formed at the same time as those in the North Atlantic? A follow-up study in 2002 showed that the increases in rubble within the South Atlantic ocean muds happened at the same time as warmer sea temperatures. A fascinating story was starting to unfold.

Before we proceed, let's just pause for breath for a moment and look at what we've covered; there's been a barrage of ideas so it's probably worth taking stock for a moment. We know from buried ocean muds that after the Eemian there were catastrophic collapses of the ice sheets around the North Atlantic that coincided with periods of major cooling. These happened periodically and would lay down a layer of rubble on the ocean floor. After each big release of bergs into the ocean, the ice sheets would then build up until another collapse took place. Reconstructions of sea level show that changes of between 10 and 35 metres coincided with the most extreme splurges of ice during Heinrich Events. The dating is not that certain but it looks like Antarctica was purging itself of ice during warming in the south while cooling was going on in the North Atlantic. It all points to massively unstable ice sheets. Could changes in sea level have been enough to cause their collapse?

When travelling around Britain you'll often read tourist bro-chures waxing lyrically about the virtues of the Gulf Stream and why it makes the western coastal resorts so unusually warm for their latitude. Usually there's a picture of something that's mis-takenly described as palm tree with a happy couple in various stages of disrobement and wild abandon. Although chucking your clothes off in the heat of the day across Britain isn't always possible, it's hard to deny some sympathy with the sentiment. You only have to look at the world climate maps to see that New York is often below freezing during winter while London is generally warmer. A major reason for this unseasonable warmth is the trop-ical water bought up from the Gulf of Mexico into the North Atlantic.

As the Gulf Stream heads north, it splits into two currents: the Canary Current which shoots off to the southeast, and the North Atlantic Drift which continues on up to the northeast. The result is warm water that's pushed up past the British Isles. As the North Atlantic Drift heads north, water from the surface evaporates, delivering the newly released heat downwind, courtesy of the winds blowing from the west; the amount of heat given off is equivalent to something like a million power stations. Continued evaporation causes an increase in the amount of sea salt within the surface water. As the warm waters head north, they keep the Arctic winter sea ice at bay but eventually get cold and salty enough to sink in the Greenland-Norwegian and Labrador seas, forming North Atlantic Deepwater. This deepwater returns south at depths of between 2 and 5 kilometres. Because temperature and salt play such a major role in the movement of this water, it's often described as thermohaline circulation (Figure 5.3).

In Antarctica, deepwater formation happens in a slightly differ-ent way to that in the North Atlantic. During winter, large amounts of sea ice form in areas of open water within the Weddell and Ross Seas. As the ice forms, heat is given off and most of the dissolved salt remains in the seawater. The result is that the water becomes dense enough to form deepwater; this flows off the

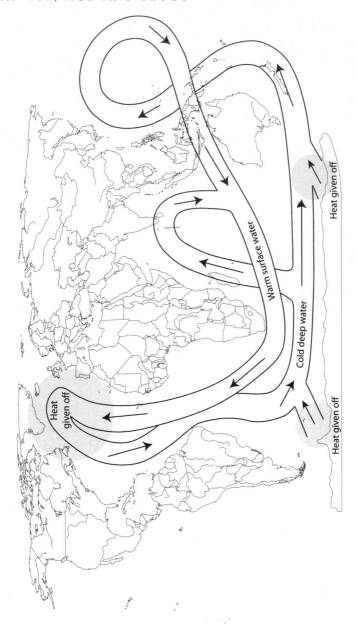

Figure 5.3 Thermohaline circulation.

continental shelf, where it meets the North Atlantic deepwater that has travelled southwards. These waters combine and then flow at depth into the Indian and Pacific oceans, where thanks to the winds and tides they eventually well up to the surface and return to the Atlantic. Because the grand scheme involves water flowing around much of the world, the whole system is often likened to a global conveyor belt.

But has this always been the case? In Harvard University during the 1960s, Henry Stommel was one of the first to suggest that differences in the temperature and saltiness of seawater might cause the ocean to circulate differently. This was real pioneering stuff. Stommel suggested that a reorganization of the world's circulation might cause climate change. It's now thought that in the past the Atlantic might indeed have circulated in three different ways. The first is just as today, with deepwater formation happening in the northern seas; the so-called warm mode. On the other hand, during colder periods there may have been some sort of 'deep' water formation, but it probably happened further south, perhaps equatorwards of the shallow sill that stretches between Greenland, Iceland and Scotland; this is sometimes called the cold mode. Alternatively, there might have been a complete shutdown of the circulation system; essentially the circulation was switched off.

In the past, it seems that the closer freshwater from melting icebergs was dumped into a region of North Atlantic deepwater formation, the more vulnerable the whole system was to shutting down. Wally Broecker at the Lamont-Doherty Earth Observatory at the University of Columbia has suggested that this is the Achilles heel of the ocean circulation system. Time and time again, climate models show that a pulse of freshwater to the right spot can cause a slowdown or shutdown in the ocean circulation, causing widespread cooling across the North Atlantic. Recent modelling work by Richard Seager and colleagues at Lamont-Doherty has shown that if the circulation system is turned off, the absence of warm waters moving north allows Arctic winter sea ice to expand; northern Europe cools by up to 20 °C while central and southern European temperatures fall by 3 to 6 °C. Ultimately, a big dose of freshwater dumped in the North Atlantic can make the region very cold.

It certainly appears that we have a mechanism to explain the cold Heinrich events in the North Atlantic. But it might not be as clear-cut as that. We know that for at least some of the Heinrich events the layers of rubble were laid down after temperatures had already started falling. This means that the bergs can't have been the original cause of the drop in temperature. The splurges of ice look like they exacerbated the cooling that was already taking place. So what caused the initial cooling?

A potential solution to this apparent impasse might lie in the south. Recently, the most detailed and precisely dated Antarctic core has been drilled by the European Project for Ice Coring in Antarctica at Dome C. The results are giving some crucial insights. The EPICA team has produced an exquisite record which shows that warming took place in the south when cooling was happening in the north. Each time it was cold over Greenland, Antarctica warmed by about 3 °C. Thanks to the detail possible in Dome C, however, the data also show that at the moment of peak warmth in Antarctica, the temperatures suddenly ramped up over Greenland and in the North Atlantic. In contrast, when it was cold in the north, the southern temperatures increased gradually. Amazingly, this pattern seems to hold over millennia. No matter how small the cooling in the North Atlantic, there was warming in the south. What could have caused such an extraordinary pattern?

Thomas Stocker at the University of Bern and Sigfús Johnsen at the University of Copenhagen have likened the Southern Ocean to one big heat reservoir. If deepwater formation is reduced or shut off in the north, the warm water from the tropics no longer heads towards the North Atlantic. Instead, the warm water relocates south, where the Southern Ocean gradually absorbs the extra heat. It's a bit like a seesaw; when it cools in the north, the south gradually warms. The longer the deepwater formation is turned off in the North Atlantic, the longer it remains warm in the south.

These periods of warming in the south may have had an impact on the Antarctic ice. It looks like the rubble layers that were identified by Kanfoush's team in the South Atlantic were formed when the ocean was warm. If these warm events are the same as those recognized in the Dome C record then a potential scenario starts to present itself. As the North Atlantic cooled (for reasons not yet clear), the heat from the tropics would have gone south, warming the Southern Ocean. As it did so, increasing amounts of the Antarctic ice shelf would have become vulnerable to melting. Eventually a tipping point would have been reached and the ice shelves would have collapsed, dumping a large amount of ice into the Southern Ocean. The impact of this would have been felt globally and instantaneously; sea levels would have risen around the world, detaching North Atlantic ice shelves from the seabed. If the Laurentide Ice Sheet was large and unstable enough, the increase in sea level might then have been enough to cause a Heinrich event; a catastrophic collapse that would have led to an armada of bergs, cooling the North Atlantic further and covering the sea floor in massive amounts of rubble.

Modelling studies have suggested that if half of a 30-metre sea level rise came from the Laurentide Ice Sheet and the other half from Antarctica, the circulation system in the north would have crashed, while warming would have continued in the south. Other models suggest that such a large release of Antarctic freshwater would have caused the south to cool regardless of how much heat was coming from the tropics. The jury is very much out. Essentially we're still not sure how and why the ocean circulation slowed down and then started up again. It's very much a chicken or the egg situation. The only thing we do know is that when large amounts of freshwater are dumped into the North Atlantic, there can be dramatic cooling. Could it happen again?

This all raises an interesting conundrum for the future. If large parts of the Greenland Ice Sheet were to melt, isn't there a good chance the North Atlantic will cool? Surely the ocean circulation would shut down or be severely weakened? This prospect was first raised by Wally Broecker back in 1987 and is of real concern. After all, it happened in the past so could happen again. The British press often gets its knickers in a twist over this. Images of ferries crossing an iceberg-infested English Channel often grace the television and newspapers.

To test for this, Jonathan Gregory at the University of Reading and colleagues pulled together the results of 11 computer models investigating the impact of rising carbon dioxide levels on the ocean's circulation. They found that there was no sudden collapse of the circulation under future greenhouse gas scenarios. Modelling 140 years into the future and with a quadrupling of carbon dioxide levels, the circulation gradually slowed down between 10 and 50%. This might be expected to cause a cooling but Gregory and colleagues found that the high greenhouse gas levels more than compensated for any reduction in heat brought north by a weakened circulation.

One potential complication to all this is what will happen along other parts of the global conveyor belt? Remember, winds and tides bring the deepwater back to the surface. In a future greenhouse gas-rich atmosphere, the degree of windiness will probably change. One of the big unknowns, however, is whether there will be more tropical cyclones. In the western tropical Atlantic these are known as hurricanes, and they nicely churn up the sea surface, helping keep the whole ocean circulation system going. Part of the problem for predicting what they'll do in the future is that it's not certain what the natural variability is. Tropical cyclones happen quickly, hit relatively small areas and often follow different routes to one another. Historical records are generally short and incomplete.

What we need is a continuous record of what tropical cyclones were doing in the past. One possibility is to look in coastal locations such as lagoons and beaches. Layers of rubble are often out of place in these contexts and probably signify that a storm threw the stuff up on land. But these layers really only represent the big storms. What about smaller hurricanes? Fortunately, because of their sheer size, the isotopic makeup of the rain from tropical cyclones is unusual compared with the stuff that falls most of the time. As this water seeps down through the soil, it might encounter a cave and help form a stalactite or stalagmite, preserving the queer isotopic signal. Recently, Amy Frappier at Boston College has been looking at stals in central America and has found unusual spikes in their isotopic content that match historically known hurricanes. In Australia, similar work by Jonathan Nott at James Cook University has extended the record of west Pacific tropical cyclones. The results hint there have been distinct periods when there were more and fewer storms. Unfortunately, however, there is still some way to go before we know what the natural variability is over the long term.

Some of the latest climate models suggest that the intensity of hurricanes will increase in the future. If there are more storms, the oceans would continue to be mixed, keeping the circulation going. So even with more freshwater swishing around in the North Atlantic it doesn't seem likely there will be a significant slowdown in deepwater formation. It doesn't look like a shutdown in the North Atlantic is likely. Europe will probably continue to warm up along with everyone else.

Chapter 6
A BELCH AND A BLAST

The end of the last ice age from 21,000 to 11,700 years ago was a tumultuous time. It wasn't a simple orderly transition from cold to warm. Instead, the world's climate veered abruptly from one extreme to the other, often doing different things in different parts of the world. Crucially for us it's also a time when we start to see people changing their behaviour in response to what the climate was doing.

I have to admit I've got a bit of a soft spot for this period; it was the topic of my PhD research. As a student I was based at Royal Holloway, a college of the University of London, just outside Windsor in the UK. My main supervisor was John Lowe, who was heading up a large research project investigating changes in climate and the environment since the ice age. Several of John's colleagues also had big research teams working on different periods of climate change. The result was a department full of students and other researchers. Intellectually it was a fantastic time. There were always ideas bouncing around; any newly published research paper would suddenly become the focus of discussion over tea and coffee. We'd help each other on fieldwork and chat through our findings. It was always fun trying to outline your latest ideas over beers. I learnt a heck of a lot.

I was keen to work on any site that might hold a clue to what happened to the British Isles after the ice retreated from the Last Glacial Maximum. It didn't matter where: a river bank in south-western Ireland; an old brick works in southern Scotland; a gravel pit in Yorkshire; a golf course in Cumbria; an open-cast coal mine in southern Wales. Although these might sound like an eclectic mix of locations, they all had one thing in common: they were all old lakes that had formed after the ice had melted away. Over time, the lakes had filled in with sediments, preserving a record of the changing environment and climate. By the time we'd got to them, most of the lakes had long since filled in.

In most cases we'd have to core a site with a bit of kit that looked like Heath Robinson had designed it. The key thing was to get several metres down so as to reach the sediments that had accumulated immediately after the last ice age. Most corers are made up of a thin-walled aluminium tube, usually a metre or so long, and look like an anti-tank missile launcher. As the corer is pushed into the ground, it captures the sediments below the surface, allowing you to analyze them back in the lab. Peats are easy; lake sediments are hard work. Thinking back, the exercise probably did look rather bizarre when seen from afar; there'd usually be a group of three to four of us, hanging off and swearing at something that looked like it should be banned under any reasonable arms treaty. In spite of all the hard work, it was good fun and a great way to try to understand how the world works. A persistent problem, however, was that the sampling tubes are usually only a few centimetres across, so we had to think carefully before running any analysis on the precious little sediment we had (corers can be made larger but are then nigh on impossible to get very deep). Fortunately a few of the sites I worked on as a research student were being eroded out from riverbanks or cliff faces, giving us lots more material to work on.

Assuming we weren't suffering from lumbago after coring, we'd often see the same distinctive suite of sediments. At the bottom of the sequences would be a thick unit of sand and gravel. This often had very little else in it and represented a time when the landscape was barren, swept clean of most life by the ice. Some time after the ice had retreated, life appears to have returned in abundance. Conditions became warm enough that plants and animals were able to colonize the landscape and live in the lake waters. The sediments shout this out loud and clear, with a clear switch to rich brown lake muds. But when it looked like the good times were here to stay, the environment dramatically changed. The organic-rich muds suddenly changed to grey clays and silts, suggesting that it had got a lot colder and the vegetation cover had been drastically reduced. Similar changes have been seen in lake sediments across the northwest of Europe. The question was how much of these changes reflected climate.

Much like the ocean cores we've already looked at, lake sediments contain a lot of different fossils that can be used to reconstruct the environment and climate in great detail. Probably one of the best known is pollen. Most pollen is designed to travel on the wind and vast amounts are often produced in the hope that some will pollinate. One of the reasons why pollen is such a curse to hayfever sufferers is that many plants produce prodigious amounts of it; a single ragweed can turn out a million grains a day, while a branch of marijuana can release more than 500 million grains. Most of these never get to do the job they were intended for; otherwise we'd be up to our necks in cannabis. Instead, there's a huge amount of waste, with the vast majority of pollen grains falling by the wayside. Fortunately, this can be turned to our advantage. Some of the pollen ends up on lake or peat surfaces, eventually becoming locked within the sediments and buried underground. Although most grains are smaller than 0.01 mm across, their shapes can vary enormously. The result is that when the different pollen grains are extracted from the sediments it's possible to identify most to at least the family or genus level: for instance, hazel trees produce pollen grains that look like samosas, while pine looks like the outline of a well-known cartoon character in the shape of a mouse. By sampling layer by layer down through a core and counting around 300 pollen grains at a time, it's possible to get a good idea of what was growing in the immediate area when the sediments were being laid down. If you see a samosa you can tick the box for hazel; the silhouette of a popular mouse-shaped cartoon character, a pine. If enough levels are analyzed through a core, changes to the vegetation in a landscape can be investigated.

In the early 20th century, Norwegian geologist Lennart von Post did some of the earliest work on deciphering what fossil pollen was saying. A formidable man in stature, he was the first to count different pollen grains within Swedish peat sequences. In 1916 he surprised his Scandinavian colleagues when he described for the first time how pollen might be used to reconstruct past environments. Because temperature and moisture play a significant role in deciding where plants can and can't grow, it didn't take a big leap of faith to interpret some of the vegetation changes as climate.

When European scientists started looking at pollen preserved in lake sediments deposited after the last ice age, they found that there had been big shifts in the vegetation. Most critically, the rich organic muds contained pollen grains that indicated trees were common across western Europe at this time. This clearly showed that the organic muds were formed during a period of warmth now known as the Lateglacial Interstadial. In contrast, the pollen types were dramatically different within the grey clays and silts. The forest cover crashed, to be replaced by a totally different suite of plants. Arctic species became the top dog. These included *Dryas octopetala*, a low-lying shrub that's common today in cold environments. The conclusion was there had been a big change in temperature. Because there was so much *Dryas octopetala* in the period represented by the silts and clays, this interval became known as the Younger Dryas Stadial. It looked like there had been a return to ice age conditions. In Britain, there seemed to be a long-term increase in birch pollen immediately before this, suggesting it had got progressively warmer until the onset of the Younger Dryas. Were these just local changes?

An obvious place to look was in Greenland. If the same changes could be seen in the ice cores it would show these were of regional significance. One of the first records from Greenland was actually taken from a site known as Camp Century, 'the city under the ice'. This bizarre place was a military installation in the far northwest of Greenland. During the Cold War, Greenland's military significance became apparent to the USA. In 1958, in fear of a Soviet military strike, Camp Century was constructed within the ice sheet so that a future counterattack might be made. The ultimate plan appears to have been to construct a vast military site, holding up to 600 nuclear missiles that could be moved around without fear of detection via a 4,000 kilometre underground railway system. Camp Century was the first phase to see whether such a scheme was possible. It was strictly men-only, with up to 250 individuals staying in an underground camp that did not want for anything. In addition to the accommodation there were a cinema, library, church, chemist, gym, barber-shop and research labs, all accessed off a main street over 330 metres long; electricity was provided by a nuclear reactor. As we know, however, ice flows. A team of 50 men in the camp were dedicated solely to keeping the tunnels clear. In spite of this, attempts to build and operate a railway failed spectacularly. By 1966 it was apparent that the movement of ice was too great. The whole place was closed and the military plans shelved.

One good thing did come from all this. Towards the end of Camp Century's life, the US Army Cold Regions Research and Engineering Laboratory succeeded in drilling a core from here down to 1390 metres. The ice was analyzed by a Danish and American team led by Willi Dansgaard and the results showed that even this far north, pronounced warming and cooling had occurred. It all seemed to tie in with what the European lake sediments were saying. There was still no real way of knowing what the temperature had done, however; although the Camp Century reconstruction was based on the oxygen isotopes in the ice it wasn't calibrated against temperature. Although it's good fun saying it got colder or warmer, relative to what and by how much?

Fortunately, it's not only pollen grains that are preserved in ancient lake sediments. In the late 1960s, Russell Coope almost single-handedly pioneered one of the great methods for reconstructing past changes in temperature using bits and pieces of fossil beetle remains. Of all the different types of insects, beetles are the most abundant and diverse. More than 300,000 species have been described and as such make up around a quarter of all known living things. Not only are they found around the world, but many species also have a strict preference for a particular environment. If you can identify the species from the fragments of fossil exoskeleton in the lake sediments it's possible to describe the environment in which they lived; rather wonderfully, one of the best pointers for identifying beetles are the male genitalia which are distinctly different at species level. Some species will give an indication of whether there was flowing water; others dry, sandy soils. Dung beetles indicate mammals had been about.

The more exciting thing from a climate viewpoint is that many beetle species only thrive within a relatively narrow temperature range. With other colleagues, Russell Coope has been able to use this information to show how temperatures have changed in the past. As early as 1969, he was showing that the highest temperatures took place at the beginning of the Lateglacial Interstadial and not at the end, as Camp Century and the European pollen suggested. Since this early work, Coope's results have been vindicated, as virtually all reconstructions – including those from the more recent Greenland ice cores – show that the warming was greatest at the start of the interstadial. Importantly, the beetle results also highlighted one of the dangers of lining up changes in sediments to temperature. It now looks like the early warming happened within the lowermost sands and gravels and not higher up when the organic muds were first laid down; the implication is that the beetles responded quickly to the temperature change while most other animals and plants took their time to migrate back into the landscape. In terms of actual temperature, the beetles show that 14,700 years ago the summer temperatures in southern England shot up by around 7 °C to 17 °C. Things then seemed to gradually cool down across the North Atlantic, reaching a low during the Younger Dryas around 12,900 years ago. The summers in southern England dropped back down to 10 °C; a return to ice age conditions.

Although a Younger Dryas summer temperature of 10 °C in southern England doesn't sound too cold, it's equivalent to today's northern coast of Siberia. At these temperatures, Brighton wouldn't have been the popular seaside resort it is today. Temperatures might have been even worse than this. If it gets too cold, many species will go to sleep, die or (if able) up sticks and move. This is a bit of headache when trying to work out what the temperature was doing in the past. Remember, this temperature reconstruction of 10 °C comes from a slab of sediment that might represent a couple of hundred years; we don't know whether this was an average or only happened once in a blue moon. We can only go with what's preserved in the mud and hope this snapshot of time is representative. It gets even worse for trying to work out what happened during winter. Living things don't do much in the coldest part of the year. If we want to see what the winter temperatures were doing, we have to go out into the landscape and look elsewhere.

As temperatures plummeted during the Younger Dryas, icy conditions returned to a large swathe of northern Europe for some 1200 years. A mini ice age had returned and frozen ground was once again the norm. But the ground didn't just sit there and freeze; it heaved. All sorts of things took place at and below the surface. In some regions, the ground cracked open, allowing water to flow in and then freeze; as each winter passed, these wedges got larger and larger as the newly formed ice pushed down and out. If the conditions are right, ice wedges can become huge, measuring some 10 metres deep and 3 metres across. The key thing is that they can only form when there is an enduring amount of permafrost. If it's too warm, they just don't form. In other spots, ice under the surface swelled up, forming small domed hills – known as pingos – across the landscape. These things can be enormous, measuring up to 70 metres high and 2 kilometres wide. Because pingos found in today's Arctic grow at the rate of a few centimeters a year, the size of those from the Younger Dryas indicate it must have been cold for a heck of a long time. It couldn't last. When the permafrost melted, the ice wedges filled in with sediment and the pingos collapsed, forming big holes in the ground. Just finding these in the landscape shows there must have been a continuous permafrost during the Younger Dryas. Plotted up on a map, these features seem to have been common across northern Ireland, Britain and on into Scandinavia. Average temperatures must have been about –8 °C, suggesting the winters were as low as –20 °C. It was damn cold.

When the Last Glacial Maximum ended, the British Ice Sheet went into steep decline. By the end of the Lateglacial Interstadial, there may have only been a few small pockets of ice left in the Highlands of Scotland. With the onset of cold conditions during the Younger Dryas, the ice had another chance to wreak havoc. It built up again in the western Highlands of Scotland. If you've watched Mel Gibson's *Braveheart*, you'll have seen William Wallace hiding from the English on Rannoch Moor, just north of Glasgow. Although it's a beautiful place when the Sun is out, most of the time it's cold, damp and wind swept. It's an ideal place to hide from the English and build up an army if you're so inclined. It's also a great place for a nascent ice cap. The ice built up here and advanced down as far as Loch Lomond.

Unlike the big ice sheets we saw during the Last Glacial Maximum, it's a lot easier to use the smaller Younger Dryas ice caps and glaciers to get a handle on climate. Modern studies on slabs of ice show that summer temperature and the amount of snow they receive control their size. Essentially if relatively more snow builds up at the top of the glacier than is lost at the bottom, the ice surges forward. If the opposite happens, the glacier falls back. If the losses and gains are around the same, the glacier stays where it is and builds up a moraine.

Studying today's glaciers shows they can be split up into two parts: an upper area, where more snow builds up than melts over a year, and a lower zone, where the opposite happens. Roughly speaking, when the ice is neither advancing nor retreating, the area of accumulating snow makes up somewhere between 60 and 65% of a glacier's surface. The practical upshot of all this is you can take a map, walk out into the back of beyond, mark up the location of moraines in a landscape and use this to reconstruct an extinct river of ice. Because we know from the beetles what the temperature was doing during the Younger Dryas, it's possible to use these former glaciers to work out the amount of snowfall.

Looking across the British Isles, the concentration of Younger Dryas ice was in southwest Scotland and the Lake District. Using the known relationship between glacier size and climate, the amount of precipitation looks like it was around 80% of today's value. Importantly, the lack of ice in northwest Scotland shows that it was relatively dry in the far north. This pattern has been found across northern Europe. In Scandinavia, the ice surged forward once again, yet further north on the tiny island of Spitsbergen in the Barents Sea the glaciers were pathetically small. It all points to the band of high rainfall that this region gets today being considerably further south.

These climatic changes appear to have been driven by what was happening in the oceans. Today, much of the rain that falls over northwestern Europe comes from low-pressure systems that sweep in from the Atlantic. First described by Norwegian Vilhelm Bjerknes during the First World War, these form on the boundary between cold Arctic air heading south and warm tropical air going north; he likened them to 'fronts' after the battles being fought in Western Europe at the time. Bjerknes realized that when air with two big temperature differences clashes, the warm air rises up above the cold, creating a low-pressure system, and generating lots of clouds and rain in the process.

As we saw in the last chapter, the North Atlantic Drift currently brings warm water north. It doesn't get much further north than Norway because cold Arctic waters come south, hugging the coast of Greenland and sweeping down the east coast of Canada. This big contrast in sea temperature is known as the polar front.

Because of the big differences in ocean temperatures over the western Atlantic, a lot of low-pressure systems form here, especially during winter. Because the air generally blows to the east at these latitudes, these low pressure systems then hurtle off to Europe at about 1,000 kilometres a day, dumping rain and snow as they go. What happened during the Younger Dryas?

In 1981, Americans Bill Ruddiman and Andrew McIntyre undertook a classic piece of work to find out what was happening in the North Atlantic through this period. Taking a number of cores from across the region, they analyzed individual layers through the sediments to establish the proportion of warm- and cold-loving forams living in different parts of the North Atlantic at different times. They found that when it was warm in Europe, heat-loving forams lived in the far north, showing that the North Atlantic Drift was flowing in a similar way to today (Figure 6.1). During the Younger Dryas this all changed. Where it had once been warm, the forams were mostly cold-loving. The Arctic waters had moved southwards, pushing the polar front down to northern Spain. Cold waters would have enveloped much of Europe. Yet as Britain and its surroundings cooled, the temperatures in the tropics remained largely unchanged. The increasing temperature difference across the latitudes caused the westerly winds to shift south, taking much of the rain and snow with them.

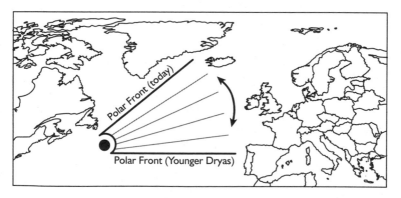

Figure 6.1 The swinging polar front.

The onset of these chilly conditions would have had a big impact on people living in the region. When it gets cold most of us throw on another sweater or ramp up the central heating. During the Younger Dryas, our ancestors seemed to have tried something else, albeit unintentionally. In 2005, Sharon Moalem and colleagues from the Mount Sinai School of Medicine in New York made a fascinating research study on the origin of Type 1 diabetes in Europe. People with this disease have an excessive amount of sugar in their blood. It's been known for some time that there is an unusual distribution of suffers in Europe, with the largest proportion at high latitudes (particularly in Scandinavia) and in Sardinia. Type 1 diabetes is less common in people of African, Asian and native American descent. The changing prevalence across Europe is uncertain. There is no obvious reason why people living in much of the Mediterranean should be less susceptible to the disease than those in northern Europe. There's certainly no obvious benefit either. Type 1 diabetes causes a range of problems, including ketoacidosis (a nasty complaint where acid can build up to dangerous levels in the blood, leading to a coma and ultimately death), heart disease, strokes, kidney failure... the list goes on. It's nothing you'd really want to go looking for. Or is it? During the Younger Dryas, much of northern Europe returned to ice age conditions. This would have been a particular problem for people who had settled in Scandinavia during the warm interstadial. People living at these latitudes would have experienced an intense evolutionary pressure. Those who could adapt to the harsh environment and deal with the intense cold would have had a far better chance of surviving.

Other species are known to load up blood sugar levels when the temperatures plummet. An excellent example is the wood frog *Rana sylvatica*, the only species of frog found north of the Arctic circle. As soon as temperatures drop, the wood frog's liver pours sugar into the blood and ultimately its organs. The result is nearly half of the frog's body can freeze; its heart, breathing and blood flow all stop. Yet after just a few minutes of thawing in springtime, all these physiological processes start up again without any apparent ill effects. Not only do higher blood sugar levels reduce the freezing point at which ice forms, but they also change the shape of the crystals, making them less likely to rupture tissues. Could high sugar levels in the blood of our north European ancestors have evolved to prevent the formation of ice crystals? This is something Moalem and colleagues have wondered out loud. When temperatures dropped during the Younger Dryas, those living at high latitudes would have been affected the most; individuals who produced more sugar in their blood would have stood a greater chance of survival. An increase in blood sugar levels at this time would certainly have been useful. Recent studies of the Y chromosome suggest that people in northern Europe might have a similar origin to those who colonized Sardinia, which would help explain today's distribution of the disease. Given the low life expectancy at this time – possibly as little as 25 years – it wouldn't have mattered that Type 1 diabetes didn't extend your life to what we'd now consider an old age. It just had to get you through enough extreme winters to reproduce and pass on your genes. If true, this would be the first example of where a disease has been shown to become more common as an adaptation to extreme climate change. It's a fascinating idea.

On the climate front, the race was on to see whether the Younger Dryas was experienced around the world. An obvious place to look was Antarctica. It was no problem measuring the different isotopes in the ice. These did seem to show there had been significant changes after the Last Glacial Maximum. The problem, as we've seen before, is that snowfall over Antarctica is much lower than Greenland. As a result it isn't possible to count layers of snow laid down each year. Early attempts at estimating the age of the ice gave widely different times for the inferred climate changes in Antarctica. A new idea was hit upon in the 1990s:

could the Greenland ice core records be used to construct a chronology for Antarctica? If this was to work, there had to be some way of linking up the cores. Fortunately, the air above us is very well mixed, meaning that for most types of gases it's possible to go from one end of the planet to the other and measure the same amount in the atmosphere. The makeup of the air bubbles trapped in the ice seemed a good place to start.

During climate shifts, there have often been big changes in the concentration of some gases within the atmosphere, many a time driven by changes in the carbon cycle. We know this is true of carbon dioxide, but because of Asian dust falling on Greenland it's not possible to reconstruct past changes from this ice sheet. What we want is a gas that can be measured in both Greenland and Antarctica. A better prospect is methane, one of the greenhouse gases we're so worried about today, and one which is not significantly influenced by dust levels.

It was originally thought that the recent geological changes in atmospheric methane was largely controlled by the amount of wetlands that covered the world's surface. Generally speaking, most of these are in Siberia and the tropics. When it gets warm, these areas receive a lot of rain and the number of wetlands increases. It's a recipe for lots of plant matter decomposing under water without the presence of oxygen. The result is prodigious amounts of methane can be produced. Recently, however, an important but hitherto unrecognized source of methane has been discovered. I briefly mentioned earlier the existence of icy carbon-rich soils in Siberia that are known as yedoma; it now seems possible that some of these may have started to melt immediately after the Last Glacial Maximum. The melting ice from these soils can cause ponds to form in depressions across the Siberian landscape, producing deep lakes that result in the ancient plant material from the yedoma also breaking down to produce methane.

Regardless of its source, the warmer the climate becomes, the more methane seems to be put into the air; when it gets cold, the reverse happens. Once released, the gas is quickly mixed around the world, allowing a small part to be entombed within the ice. This all means that a team of scientists can come along thousands of years later, extract the ice, crush it, and measure the amount of gas trapped in the bubbles. Over the past 650,000 years, Antarctica shows that methane levels almost doubled when the climate warmed up: there were 380 parts per billion of it in the air during an ice age compared to interglacial levels of up to 725 parts per billion.

Because methane only lives in the atmosphere for around 12 years, the same peaks and troughs in gas levels can be used to pre- cisely tie together the Greenland and Antarctic ice cores. Even during shorter and more abrupt periods of climate change, the methane levels seem to have shifted significantly, ideal for linking up the records. The beauty of all this is that the excellent chro- nology from Greenland can be used by the Antarctic ice cores, allowing a direct comparison to the North Atlantic region. When this was done, a fascinating pattern emerged. The Lateglacial Interstadial and Younger Dryas in the North Atlantic did not happen in Antarctica. When it was warm in the north, it was cold in the south. When it got cold in the north, it became warm in the south.

Did this relationship hold across the hemispheres? It's easy to forget that ice cores aren't necessarily representative of such vast regions; they are after all primarily recording climate near the poles. Perhaps the climate changes in Antarctica weren't experi- enced anywhere else. It's important to know. Very few people live at such high latitudes. What were conditions like closer to the Equator where most people live? A good place to look in the southern hemisphere is New Zealand.

New Zealand is one of those perfect places for climate research. An island nation comparable in size to Great Britain, it's sur- rounded by oceans. The prevailing westerly winds slam into a wall of mountains on the west coast of the South Island, forcing the air to rise and leading to huge dumps of rainfall. As a result, this area has one of the highest rainfalls recorded anywhere in the world; 7 to 8 metres a year are not unheard of. During the last ice age, huge glaciers surged out of the mountains and dropped into the valleys below. If you ever get a chance, take a cruise in Milford Sound, a fjord in the southwest of the island. Assuming you can see through all the rain, you'll find a textbook example of an ice-forged landscape: sheer cliffs mark the limits of where the once vast glacier flowed, while hanging valleys show where small glaciers fed into the main ice mass; it's a magical place. Warming at the end of the ice age caused the glaciers around the country to rapidly melt, retreating back up into the mountains where some can still be seen today. One of the best examples is just up the coast from Milford Sound at Franz Josef Glacier.

Franz Josef used to be a sleepy place on the West Coast. Now it's a bustling tourist town used as a base by thousands to visit the stunning glacier in the adjacent valley. Yet on the other side of the town there is a huge mound of dirt and rubble that most people ignore as they hurtle along the road to go to the next tourist spot. This enormous U-shaped moraine is known locally as the Waiho Loop and is large enough to be marked up on most maps. Recent modelling of how the glacier responds to climate has shown that over the long term, the Franz Josef mostly responds to temperature. To get the ice all the way out to the Waiho Loop would suggest that the temperature had dropped by somewhere between 4.1 and 4.7 °C. The question was when had this taken place?

I've been fortunate to visit the area several times and was intrigued by the Waiho Loop's age. Early work had concentrated on radiocarbon dating wood fragments that had been trapped in sediments as the ice had advanced down the valley to the Waiho Loop. The results had suggested the advance had happened at the same time as the Younger Dryas in the North Atlantic. This caused quite a stir because it suggested that the world's climate was cooling down in both hemispheres at the same time. Not everyone was so sure, however. Early on it had been argued that the samples used to date the advance of ice were contaminated.

Radiocarbon dating uses the principle that virtually every living thing is radioactive; fortunately the amounts involved are minuscule and are nothing to worry about. The method is based on the fact that a radioactive form of carbon is created in the upper atmosphere as a result of the bombardment of nitrogen gas by high-energy particles from outer space. The end product is radiocarbon – sometimes referred to as carbon-14. This quickly gets turned into carbon dioxide with the result that it can then be taken up by plants via photosynthesis. The plants get eaten and the radioactive carbon goes up through the food chain. When alive, most plants and animals have the same amount of radiocarbon as the air. As soon as they die, the radioactive carbon starts to disintegrate, turning back into nitrogen. Every 5,730 years, half of all the radioactive carbon disappears. By measuring the amount in a sample and making an assumption about how much there was in the first place, it's possible to calculate an age.

The problem with dating wood in such wet places as the West Coast of New Zealand is that as the rainwater drains through the ground it picks up all sorts of young carbon from the soil. Any wood buried in the ground can act like a sponge, soaking up some of this carbon from the water. If the wood isn't properly cleaned in the lab, the sample can seem a lot younger than it really is. In 2005 I had a chance to visit the Waiho Loop with a visiting Dutch student called Niek de Jonge and we used this opportunity to collect some new samples of wood for dating. I wanted to avoid the issue of contamination so gave the wood samples a strict chemical treatment to clean them up. I submitted the samples for dating and immediately went on tenterhooks. Several weeks later the ages came back; the samples were 13,100 years old. There had been a contamination problem with the original study. Importantly, it appears the Franz Josef glacier had surged down the valley at the same time as there was cooling in Antarctica and warming across the North Atlantic. Recent work on ocean muds off Australia and New Zealand show the sea in this region was also cold at the same time. The

Younger Dryas was not global. The south Pacific was dancing to the same tune as Antarctica.

The ice cores from Greenland and Antarctica show the same sort of seesawing pattern in temperature that we saw with the Heinrich events. A change in sea level, however, doesn't seem a likely cause of the different trends seen across the world at this time. Immediately after the Last Glacial Maximum, a large amount of the world's ice sheets were still intact. As a result, vast areas of continental shelf were still above water. One of the largest was off southeast Asia. Here the islands of Borneo, Sumatra, Java and Bali all joined the continent, forming the enormous Sunda Shelf of 1.8 million square kilometres; this is equivalent to the size of Mexico. The latest work on sea level change shows that much of this area flooded during a single large rise of 16 metres over just 300 years. The dating, however, shows this happened around 14,300 years ago, far too late to have caused the Lateglacial cooling around Antarctica and too early to be the source of the Younger Dryas in the North Atlantic. It had to have been something else.

Carbon dioxide levels rapidly increased in the atmosphere after the Last Glacial Maximum. As mentioned in the last chapter, the Southern Ocean is a major player in controlling the uptake and release of carbon dioxide from the deep sea and this seems to have closely tracked the temperature changes seen in the Antarctic ice cores. When it was warming up, more carbon dioxide was being belched from the Southern Ocean. Although this put the world on a fast track to interglacial levels of greenhouse gases, these changes were being driven by a warming world rather than the other way round. Carbon dioxide can't have been the cause of the temperature trends.

A possible solution lies in the north. At this time the Laurentide Ice Sheet continued to cover a large part of North America. As conditions began to warm over the North Atlantic, the ice began to melt. Not all of the freshwater flowed into the sea, however; a large amount ponded up inland, building a vast lake, known as Lake Agassiz, over what is now central north Canada. Today's Great Lakes to the south are just small fry compared to the size of Agassiz. Over its 4,000-year lifespan, the lake greatly changed in size as it periodically lost huge volumes of water. An army of researchers have walked over this landscape, mapping and dating the old beaches, spits and escarpments left behind. During the Lateglacial, Agassiz looks like it reached a size of 300,000 square kilometres; it was larger in area than the United Kingdom. Originally it was thought that just before the Younger Dryas, Lake Agassiz managed to find a weak spot in the ice and drain out via the east coast, capping the North Atlantic with freshwater, shutting down deepwater formation and disrupting the thermohaline circulation.

One problem with this idea is there is no clear evidence of a big pulse of freshwater into the western Atlantic at this time; there's not even a significant layer of rubble that we saw with the Heinrich events. The amount of freshwater released can't have been that large either because there wasn't a big jump in sea level at this time. Whatever happened had to be more subtle than this.

When researchers have gone back and redated possible escape routes for Lake Agassiz along the east coast, the ages are all too young for it to be the cause of the Younger Dryas. An alternative has been suggested. Jim Teller at the University of Manitoba and colleagues have proposed that Lake Agassiz may have poured into the Arctic from the northwest. A computer simulation of how the North American ice evolved over time supports this idea. It is possible the freshwater took this route and continued round the back of Greenland before flowing into the North Atlantic. If so, this release of water would have bought tonnes of sea ice with it from the Arctic, forming a barrier to deepwater formation. Inadvertently it would have been a remarkably effective way of shutting down the circulation in the north. Compared to Heinrich events there wasn't a sea level rise of up to 35 metres. In fact there was hardly any noticeable change at all. Yet deepwater formation took over a thousand years to recover, plummeting the region into a mini ice age. It just goes to show that it's not what you've got that matters, it's what you do with it that counts.

The transition from the end of the last ice age into the current interglacial was far from smooth. A whole host of different things were going on in different parts of the world. At one level, the changes that happened are very similar to those that took place during Heinrich events. But rather than an armada of icebergs streaming off the North American east coast, it seems likely that a pulse of freshwater from the melting Laurentide ice poured into the Arctic, shutting down deepwater formation and cooling the North Atlantic for over a millennium.

Excitingly this might not be the end of the story. A new idea has just come out of left field. A team of scientists and archaeologists led by Richard Firestone at the Lawrence Berkeley National Laboratory in California have got together and argued that a comet more than 4 kilometres across may have exploded over North America at the beginning of the Younger Dryas.

The people who inhabited North America immediately before the Younger Dryas are known as Clovis; named after the New Mexico town where their distinctive flute-shaped stone tools were first discovered. These people appear to have prospered immediately before the Younger Dryas, hunting weird and exotic creatures in North America that included mammoths. They were clearly able hunters and apparently capable of dealing with climate change; the oldest well in the USA is Clovis-built, showing these people could access water when it got dry. Yet around the beginning of the Younger Dryas, the Clovis and mammoths disappeared; a people bearing different stone tools and hunting smaller mammals, such as bison, replaced them. Before 2007, the big toss-up was whether it was humans or climate that wiped out the mammoths. Now we can throw into the heady mix a cometary impact.

The team appear to have found some 25 sites containing little metal balls, shocked quartz, exotic elements – such as iridium – and black mats of carbon that all suggest a comet exploded over North America. The critical thing is all these things have been discovered at levels in the ground equivalent in time to when the Clovis and mammoths disappeared and the Younger Dryas began. Perhaps surprisingly, there is some indirect support: there is little evidence of anything like the Younger Dryas happening at the end of earlier ice ages; when it had warmed up in the past it stayed warm. This suggests something unusual happened during the Younger Dryas. There's no obvious evidence of an impact crater, suggesting that if real, the comet either exploded in the air – spreading material all over the continent – or the ice sheet was large enough to absorb the hit. If so, was the impact big enough to melt enough ice to let Lake Agassiz escape – possibly into the Arctic – and cause the Younger Dryas?

It's too early to say whether a comet really did cause the Younger Dryas. It's an absorbing idea but only time will tell. If true, it's a nice (though not very comforting) example of where a freak event can cause devastating climate change. It's something we can do very little about. Yet even here there is a lesson. The climate changes that happened at this time were clearly felt by our ancestors. Ultimately, the Clovis culture couldn't continue in the changed climate and environment of North America. In Europe, our ancestors retreated to sunnier climes and arguably may have begun to suffer more from Type 1 diabetes. As we continue with our story, we'll see other examples of how people in the past have responded to climate change.

Chapter 7
MELTDOWN

Around 18,000 years ago, the climate pendulum swung in favour of warmer times. The Earth continued to orbit the Sun as it had done through the ice age, but summers in the northern hemisphere started to see a little more warmth. At first, it was just a glimmer; a little extra warmth here and there. Over the next few thousand years, however, the heat from the Sun continued to rise inexorably over the high latitudes of the northern hemisphere. The effect was striking. Temperatures rose 4° to 7 °C and many of the ice sheets in the north began to melt. We now know the increasing warmth from the Sun wasn't enough in itself to cause all of this massive warming; there were other feedbacks that exaggerated the effect.

Up till now, we've had to rely on just a handful of natural archives that contain a record of climate. One of the inherent problems with investigating the distant past is that few records are available for study. When we look outside we can see trees, lakes, ice and oceans; all contain some record of what is happening now. But almost inevitably on a busy planet, ancient archives get destroyed with the ravages of time: glaciers bulldoze landscapes; floods erode lake basins; plant and animal remains decompose. In many ways, it's a wonder we have anything to look at all. Fortunately, as the world started to warm up at the end of the last ice age, many of these problems disappear and we start to find a lot more records surviving through to today. As a result there's a lot more information on what happened in the recent past. And as we get closer to the present day, we can date these records more precisely and tease apart what happened when.

In 1993, one of the most influential papers of my research career was published. An American team, led by Richard Alley at Pennsylvania State University, looked at the end of the Younger Dryas at Greenland's Summit. Because it was possible to count the layers of ice year by year from the surface, the team was able to get a detailed idea of what was happening to the climate in this part of the world. They found that the Younger Dryas didn't end gradually 11,700 years ago. Alley and colleagues showed that the amount of snow falling over Greenland doubled in three years. If this wasn't staggering enough, most of the increase happened in just one year. This wasn't slow, this was really abrupt. There was no gradual change for anyone living in the North Atlantic region. One year you were in an ice age, the next an interglacial. I'm still blown away by this. As the northern hemisphere slowly received more heat from the Sun, a tipping point must have been reached and the climate system lurched into a new gear. It wasn't something I had really comprehended before.

This change in the Greenland ice 11,700 years ago marked the start of the interglacial we're still living in; an interglacial that has the distinction of being defined as a geological period in its own right: the Holocene. But did it start at the same time around the world? The work by Alley and colleagues was followed up by a clever piece of scientific work led by Jeffrey Severinghaus at the University of Rhode Island. The following may take a little explaining, but it's well worth understanding because from just a single core taken at the Summit of Greenland it's been possible to disentangle the timing of climate change across the northern hemisphere.

Imagine a world where for an afternoon not a breath of wind blew and the air remained dead still. In a world like this, any heavy gases would sink to the ground while the light ones would remain aloft for longer. Although this is highly unlikely to happen, something similar does take place in the ice. Let's look at nitrogen, the most abundant gas in the atmosphere. Just as with carbon and oxygen, it has two common isotopes, in this case nitrogen-14 and nitrogen-15. We know that in the upper 50 to 80 metres of the Greenland Ice Sheet, the snow is halfway to being ice; something known as firn. Because of gravity, nitrogen gas containing the heavier isotope will slowly sink to the bottom of the firn. This separation is sped up if the surface of the ice sheet suddenly warms. When this happens, the lighter gas in the firn moves towards the warmer, upper part of the ice sheet, while the heavy gases scamper towards the colder, deeper section; a process known as thermal fractionation.

Over time, the heat at the surface will move down through the ice. If this happened at the same speed as the gases were separating there wouldn't be too much to write home about; essentially the proportion of nitrogen-14 and nitrogen-15 in the icy gas bubbles would remain the same. But the wave of heat takes about a tenth of the time the gas isotopes do to respond. The result is that when warming starts, any nitrogen gas at the bottom of the firn will soon start to have more of the heavier nitrogen than normal. By the time the surface heat has got down to the bottom of the firn, the bubble is safely locked away in the ice with more of the heavier isotope; once it's entombed, the proportion of isotopes is fixed. When Severinghaus and colleagues went to measure the proportion of different nitrogen isotopes in the gas bubbles they found a pronounced spike marking the warming at the beginning of the Holocene. The critical question was what was the rest of the world doing at this time?

Just as nitrogen is being trapped in the Greenland bubbles, so too are other gases. This includes methane. We saw in the last chapter that the major sources of methane are in Siberia and the tropics. Severinghaus' team found low methane levels in the same bubbles that showed Greenland was warming. It was only in the next sample, equivalent to some 30 years later, that methane began to increase. These results revealed that methane as a greenhouse gas was not a direct cause of the warming at this time; rather it was being produced by the change in climate. Importantly, it pointed to the tropical and Siberian regions getting wetter and warmer after the North Atlantic had already changed. In spite of the vast amounts of time involved, here was clear evidence that climate change wasn't happening at the same time everywhere. The most likely explanation was that the thermohaline ocean circulation had started up again after the Younger Dryas, bringing more heat to the North Atlantic; it was a few decades later that the wetland regions started to respond.

As the Holocene started, large swathes of the north continued to warm. As the ice melted across the Arctic, the local flora and fauna began to respond. Darrell Kaufman at the Northern Arizona University has led a huge group of scientists looking at this crucial period across the Arctic. In 2004, the team reported an exhaustive study of 140 sites north of 60° to look at what the climate had done: tree lines were reconstructed from fossil stumps of wood and shown to be above today's level; fossil plant and insect remains of southerly-living species were discovered in the far north; and offshore, warm water-loving seashells were found to have lived across large parts of the region. On the Seward Peninsula in Alaska, ancient timbers dating to between 10,500 and 9,500 years ago show that they had been gnawed by beavers. It all pointed to plants and animals living further north than they do today. Many of these observations could be converted into summer temperatures. They showed that this part of the world during the early Holocene was around 2 °C warmer than in the 20th century.

Although this might sound like a nice simple story, it wasn't a global picture. The peak in temperature didn't happen at the same time around the world. The high temperatures Kaufmann and colleagues found for the Arctic were caused by more summer heat from the Sun falling in this region. In Alaska and northwest Canada, for instance, the main warm phase happened between 11,000 and 9,000 years ago; in northeast Canada, the peak in warmth was delayed several thousand years because of the cooling effect from the remaining Laurentide Ice Sheet. In northern Europe, the peak in summer warmth happened some time later, between 7,000 and 5,000 years ago, as the heating from the Sun moved to favour this region. The timing of maximum warmth varied over the planet's surface.

We know that the early Holocene warming was fatal to the ice sheets left over from the Last Glacial Maximum. What's less clear is what happened to sea ice in the Arctic. Unlike glacial ice on land, sea ice is pretty much just frozen water. There's nothing else really to it. When it melts, it leaves no physical evidence of where it has been over the sea; there's no debris or rubble littering the sea floor. There's also no detectable change in sea level because to float sea ice has to displace enough water to support its weight. Archimedes supposedly discovered this effect while taking the most famous bath of all time. Realizing that his body displaced a volume of water equal to his weight, he allegedly leapt out of the bath and ran through the streets of the ancient Greek city of Syracuse, proclaiming 'Eureka', or in English, 'I've found it'. It's not entirely clear whether this is all true, but the premise is right. It's the same reason that ice cubes melting in a generous gin and tonic don't cause the glass to overflow.

To understand what the Arctic sea ice was doing in the past, we need to look elsewhere. Canadian Art Dyke has developed an ingenious approach. It involves using the fossil remains of a living organism, in this case the bowhead whale. These graceful creatures have the largest head and mouth in the entire animal kingdom, making up one-third of their body. Crucially, each year they follow the expansion and contraction of the sea ice front. As temperatures plummet through the winter, the whales move out as the sea ice increases in area, reaching most northern coastlines by March. But as summer kicks in, the sea ice coverage falls back so that by September the ice and the whales have retreated towards the North Pole. During the summer, two different populations, the Bering Sea and Davis Strait bowhead whales, converge on the Canadian Archipelago. Today they can't meet because of a continuous sea ice barrier in the Northwest Passage.

Dyke and his team have now found over 1,200 bowhead whale remains of which more than 500 have been radiocarbon dated. By examining where the fossil remains were discovered, the extent of the sea ice can be tracked over time. The results leave no doubt there was a lot less sea ice at the beginning of the Holocene. The ages show that the Bering Sea whales were the first to reach the Canadian Archipelago, about 11,500 years ago, entering via the Beaufort Sea. As the ice continued to melt, the ice barrier that keeps the two populations apart today looks like it melted some time around 10,700 years ago. The result was the Bering Sea and Davis Strait whales were able to mingle until around 8,900 years ago, when the sea ice grew out once again. The year-round ice barrier has remained in the Northwest Passage ever since.

As the icy wastes fell back, the seas of the world rose. At the height of the last ice age, 21,000 years ago, sea level was 120 metres below where it is today. By 7,000 years ago, most of the ice had melted. Vast amounts of water were returned to the sea, raising levels to within 4 metres of what we'd recognize today. But dumping more water into the ocean wasn't the only cause of sea level change.

When the ice melted, a huge burden was removed from the land. Areas that had been under the ice sheets started to correct themselves. If large enough, ice sheets have the same effect as Archimedes' body. The larger and longer one remains on the surface, the deeper the land is pushed. As the land sinks, nearby areas adjust by lifting up. But when the ice disappears, the surface shifts to a new equilibrium: the land formerly covered by ice rises while adjoining areas sink (Figure 7.1). Although this seems simple enough, the readjustment doesn't happen overnight.

In 1731, the great Anders Celsius – he of the temperature scale fame – heard tales of a strange rock on the small island of

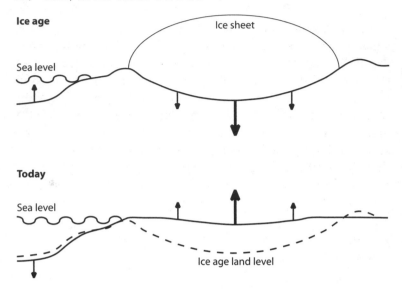

Figure 7.1 Rebounding land after the ice.

Lovgrund off the east Swedish coast. The old fishermen insisted that when they were younger, seals would obligingly climb up onto the rock, allowing the locals to shoot them. Now it seemed the seals couldn't climb the rock because it was too high relative to the water. The fishermen weren't that happy, though presumably the seals were rather pleased with the new state of affairs. Celsius clearly wasn't someone who took old fishermen's tales at face value and promptly set off for the island to investigate. He cut a notch in the rock at the level of the sea. Sure enough, over the years, the rock continued to put distance between itself and the water. Celsius argued that the land was rising up out of the ocean, though he wasn't sure why. We now know that Celsius was on the money. Relative to the sea, Scandinavia is rising. The effect is so large that although global warming is currently driving sea levels up by 3 millimetres a year, it's completely masked by the land bouncing back at 10 millimetres a year. But Celsius and the old fishermen were not the first to notice this phenomenon. As early as 1491, the inhabitants of the Swedish town of Östhammar had petitioned their king and archbishop to move closer to the sea because their boats could no longer get into the harbour.

As the ice retreated, tectonic changes were also put in train. In the 1970s, a 300 metre high scar was discovered in a headwall off the west coast of Norway. This was the source of a devastating slip known as the Storegga Slide. As the base of the wall collapsed, around 3000 cubic kilometres of rock and sediment launched itself into the Norway Basin in one big catastrophic pulse. It's the largest known slide to have ever happened under the sea. Some 95,000 square kilometres of the ocean floor was covered in debris, reaching halfway to Greenland. But the whole thing couldn't have happened without an ice age thrown into the mix.

As the Scandinavian Ice Sheet waxed and waned, huge amounts of sediment were bought out from the interior and dumped on the edge of the ice. Normally, when sediment is dumped outside a glacier or ice sheet, the water is swept away. But at Storegga, the sediment looks like it built up so fast that the water became trapped. When sediment becomes soaked in water it's only a matter of time before the whole thing can float on its own accord: it just needs a slight push. With the Storegga Slide it looks like the encouraging nudge came from the rebounding Scandinavian landmass. As the land began to rise, the bedrock was forced to adjust. A series of powerful earthquakes took place and one or more of these was probably enough to kick start the slide. Once the slip got started there was no stopping it.

It's hard to get a precise date for when all of this happened. Cores have been taken across the Storegga Slide but it's only been possible to find and date forams in the ocean muds that are above and below the layer of debris. Fortunately there are other ways to get a handle on the age. Not all the action was confined to under the waves. Thick layers of sand and gravel have been found in lakes and peat bogs up and down the coastline, indicating that the slide of debris was large enough to form a devastating tsunami. The deadly waves would have struck the nearby coastal areas within minutes. Along the Norwegian coast, the tsunami waves must have been 10 to 12 metres high when they struck land. Further afield, the tsunami would have taken several hours to arrive. On the east coast of Greenland it would have been much smaller that which struck Norway, yet even this far from the Norwegian shelf, the tsunami was powerful enough to create a 3 metre high wave.

This didn't all happen with no one watching. Coastal settlements would have been devastated. Even as far afield as Inverness in Scotland there is evidence that the tsunami took out populations of people. Overall it looks like some 600 kilometres of the United Kingdom coastline was hit broadside by the tsunami. On the west Norwegian island of Fjørtoft, it appears that a Stone Age settlement was buried under 1.2 metres of sand and gravel when the tsunami struck. Part of what was buried was a building site. This is great for dating. With radiocarbon, you really want to use plant material that was alive just before the event that's being dated. With a building site, there's a good chance some of the material was prepared shortly before it was to be used for construction. The youngest age from Fjørtoft suggests the tsunami struck some time between 8,400 and 8,200 years ago.

In North America, the Laurentide Ice Sheet persisted longer than most, although by 9,000 years ago it was a shadow of its former self. With the ever-changing way that the Earth orbits the Sun, the early Holocene had been particularly warm and the ice was melting away. It might have been thought to have done its worst; after all, the days of Heinrich events were just a distant memory. But the Laurentide Ice Sheet had one last big card to play. As the ice shrank back to the Hudson Bay, it continued to pour freshwater into Lake Agassiz along its southern margin. It was only a matter of time before the ice barrier gave way.

Sometime between 8,700 and 8,200 years ago the dam finally broke. It was a cataclysmic event. A torrent of freshwater and icebergs flooded into the North Atlantic. It was the largest single pulse of freshwater over the past 100,000 years. The impact was virtually instantaneous. In the North Atlantic, ocean circulation slowed down, stopping warm tropical waters from getting north. Temperatures plummeted and stayed low for nearly 200 years. The most detailed records come from the Greenland ice cores and suggest that it became 6 °C cooler over the ice sheet. Downwind, Europe also felt the crunch, with the main rain belt shifting south and temperatures dropping by nearly 2 °C. Similar things look like they happened in equatorial regions as well, with the tropical rain belts moving south. At the same time, methane levels in the air dropped 15%, suggesting wetland areas became cooler and drier.

But how much of a slowdown in ocean circulation would drive these massive changes? It's not just an academic question. The freshwater burst at this time could be a useful comparison for the future. Unlike the Heinrich events, many of the big ice sheets had melted away, and sea levels were only around 4 metres lower than they are today. It's not a far cry from here and now. If a comparable release of freshwater were delivered into the North Atlantic in the future, how much cooling would there be without today's increasing greenhouse gas levels? Unfortunately it's unclear from what's preserved at the bottom of the North Atlantic.

One way to work out how much the ocean circulation slowed down is to see whether a climate model can recreate the changes seen in Greenland and elsewhere. This is exactly what Allegra LeGrande and colleagues did at the NASA Goddard Institute for Space Studies. Taking a climate model that simulated changes in the air and ocean, they looked at dumping different amounts of freshwater at different rates into the Hudson Bay and North Atlantic. LeGrande and colleagues watched the impact of these changes and compared the results with what we know happened on the ground. The best fit happened when the amount of deep-water formation dropped by half. This was enough to cause the known cooling across the North Atlantic and the decline in tropical rainfall.

Changes in temperature and rainfall weren't the only thing that happened. Other effects were felt further afield. With all the extra water dumped in the North Atlantic, the world's sea level rose by up to 1.4 metres. Many areas on the continental shelf suddenly flooded. This included large parts of the North Sea. Recent work by a team from the University of Birmingham has used sonar to map 23,000 square kilometers of the sea floor; they've shown that it wasn't a bleak featureless landscape. Instead, a vista has been revealed that was hidden for over 8,000 years: a huge plain made up of hills, rivers, coastlines, sand banks and salt marshes. An unexplored country just below the waves, but a world away from what we know.

A couple of hundred thousand years ago in Britain, it didn't matter what sea level was doing; you could have visited any time you liked. Southern England was connected to mainland Europe by a ridge of chalk that stretched across the Dover Strait. During earlier ice ages, the White Cliffs would have been part of a land bridge that stretched all the way across to France. Each time an ice age happened, the central and western European rivers that flowed into the southern North Sea found their exit routes blocked: the Scandinavian and British Ice Sheets lay to the north while the chalk ridge was in the west. The water had nowhere to go. As people travelled back and forth across this Dover bridge, they'd have been greeted with a majestic site: a vast lake some 600 kilometres across. But as each ice age came and went, the chalk ridge weakened a bit more. Finally, sometime around 200,000 years ago, the dam gave way, the lake escaped, and in the ensuing maelstrom the English Channel was forged. Before this momentous event people were frequent visitors to Britain; now they only had the option when the sea level was low enough for the ocean floor to be exposed.

During the last ice age, people would have fled south across the old seabed, effectively abandoning Britain for warmer climes. During the warm Lateglacial Interstadial they returned. There's a whole host of burial sites and artefacts which show that people were living across large swathes of Britain at this time. This was dramatically bought home to me when I was doing my PhD. A fellow student and friend, Charlie Sheldrick, made a stupendous discovery in the lab. Our team were investigating sediments from a Yorkshire gravel quarry known as Gransmoor so as to reconstruct the past climate and environment in this part of Britain. Charlie was studying a large block of wood that had been pulled out from the site when he found a beautiful barbed antler point. Dating showed this artefact had been lost at the end of the warm part of the Lateglacial, some 13,000 years ago; a hunter-gatherer must have lodged the point in the block of wood when it was lying at the bottom of the old lake; they must have been furious at its loss.

The descendants of these hunter-gatherers weren't here to stay, however. It doesn't look like anyone bothered hanging around during the Younger Dryas. Although the sea levels were low enough to connect Britain and Ireland to mainland Europe, it was probably too cold to stay all year round. Hunting groups probably made brief visits in the summer to hunt deer and other large animals, but it's unlikely it was worth the effort to linger over the winter. As the North Atlantic warmed, however, temperatures rose across Europe. The land started to defrost and the British Isles shot up the property stakes as hunters recognized the new opportunities. It would have been easy for people to grab their few belongings and drag the family across the valley that would later become the English Channel. It was a time for moving. By around 11,100 years ago they'd arrived.

But the good times didn't last. It's unclear precisely when the North Sea and English Channel did join up, but the ages suggest around 8,300 years ago, the same time that Lake Agassiz burst. No matter what the cause, the result was final. Britain became an island and would remain so – at least for the remainder of this interglacial. Huge numbers of people must have been forced to move by the rising tides. Britain was on its own.

Although this sounds terrible, bigger changes were afoot in and around the Black Sea. On the boundary between east and west, this region has historically played a major role in trade between Asia and Europe. But back in the last ice age the Black Sea remained separate from the Mediterranean by an enormous ridge known as the Bosporus Sill. The sea was kept back, allowing a large freshwater lake to flourish. Large parts of the old lake have been mapped with sonar. The results show a world 155 metres below today's sea surface, with meandering river valleys, deltas and beaches. As the world's ice sheets started to melt, the Mediterranean converged on the sill. Eventually, the dam was breached and a cascade of seawater poured into the lake. It took over 30 years for the Black Sea to fill, flooding an area equivalent to the size of Ireland in the process.

One way to date when the Black Sea flooded is to look at what was living in the water. At one point, the shells of freshwater molluscs change to those that thrive in seawater. Radiocarbon dating these remains shows that the Black Sea flooded around 8,300 years ago, the same time as Lake Agassiz poured into the North Atlantic and Britain got separated from Europe. It looks like the collapse of the last North American ice was enough to make the Black Sea join the rest of the world. It was the straw that broke the camel's back.

Importantly, it doesn't look like the pulse of freshwater into the North Atlantic caused instant cooling. The results of LeGrande and colleagues suggest it took several decades after Lake Agassiz burst before the ocean reached its coldest temperatures; as the freshwater poured into the North Atlantic, deepwater formation gradually slowed until half its production had stopped. It appears that the collapse of Lake Agassiz and the flooding of the Black Sea happened before the coldest temperatures.

In 1997, William Ryan and Walter Pitman of the Lamont-Doherty Earth Observatory of Columbia University provoked a huge storm that is still reverberating around the corridors of academia. It had been known for some time that if you look at the archaeological evidence, farming suddenly seemed to sweep across Europe around 8,000 years ago. The team committed heresy to many archaeologists by suggesting the flooding of the Black Sea might have been the push that got farming into Europe. Could it have also been the origin of the Noah's Ark story and other flood myths in the region?

The whole question of what drove farming across Europe has perplexed archaeologists since the 1920s. The earliest evidence for the origins of Western agriculture is in the Near East and dates back to the Younger Dryas. It's here that many wild varieties of today's crops were first cultivated; wheat, barley and rye all grow wild in the region. But it's still not clear why people suddenly started domesticating plants at this time. We know that some groups had been living in the region from at least 23,000 years ago. These people used some wild varieties, but it doesn't look like there was any earlier attempt to farm them. Something changed during the Younger Dryas. Perhaps the population had started to grow? Perhaps the colder conditions were an extra stimulus to cultivate wild plants and get better yields? Perhaps it was a combination of these and other things? Regardless, the result was that by 11,500 years ago groups across the Near East were practising early farming. Slowly the new fangled technique started to get a foothold in other regions. Hunter-gathering became less popular. But once farming got across to Greece, expansion seemed to slow down (Figure 7.2).

When you take a close look at the ages for the earliest farming in Europe, an intriguing pattern emerges. The spread of agriculture out from the Near East into Turkey, Cyprus and then Greece suddenly stops around 8,300 years ago. It was suddenly not the thing to start trying. But this appears to have been a short-lived blip. By 8,200 years ago there was an explosion of farming across continental Europe; by 7,300 years ago it was widespread. By this time the British Isles was separated from the European mainland and took longer to catch on; farming didn't become popular there for another 1,600 years.

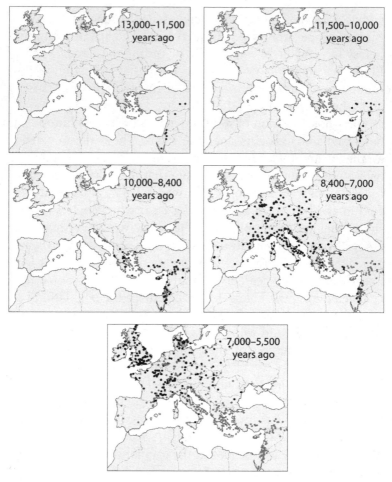

Figure 7.2 The spread of European farming.

One view is that the expansion in farming was the result of hearing about how great it was. As farming became established, word got out into the neighbouring areas. Hunter-gatherers saw the obvious benefits, it was argued, laid down their spears and joined the party: farming became the choice of most alpha one males. This must have happened to some extent, but why the collapse in farming 8,300 years ago, followed by a sudden explosion in activity across the region? Word of mouth doesn't explain all the evidence.

Colder conditions can't have been the cause for the slowdown in farming either. The Greenland ice cores show the lowest temperatures happened 8,200 years ago, yet this was when farming was becoming popular again. An alternative is the sudden flooding of large areas of the European coastline and Black Sea forced people to move on. Maybe it was early farmers living on the shores of the Black Sea. Alternatively, hunter gatherers might have been the ones who had to relocate fast; if so, they would have caused a headache for farmers living in neighbouring areas. Either way, it would have certainly seemed like the end of the world to those living near the shoreline. As the tides rose, did people move around the coastline and inland, taking farming and tales of destruction with them? It's an intriguing story. If true, it shows that it's not just abrupt shifts in temperature that can have a big effect on people. A change in sea level looks like it might have had a disproportionate impact on our ancestors, driving mass migration and cultural change. It's a disconcerting omen for the future.

These changes in our ancestors' behaviour may have had some unintentional but far-reaching consequences. Bill Ruddiman has put forward one of the most thought-provoking ideas of recent years: humans have been inadvertently influencing the climate for the past eight millennia. It's generated a huge amount of research interest. Although there's still a lot of work to do, the implications are potentially so important that it would be scandalous not to describe how this might be. If you're interested in learning more I'd strongly recommend Ruddiman's highly readable book *Plows, Plagues and Petroleum*.

As we now know it's possible to reconstruct the makeup of the atmosphere by analyzing the gas bubbles preserved in the Greenland and Antarctic ice cores. The amount of methane changed a huge amount with the end of the last ice age and culminated with a peak in atmospheric levels of 725 parts per billion between 11,700 and 8,500 years ago. After this time, methane levels fell slowly to 625 parts per billion around 5,000 years ago. Then something unusual happened: the concentration began to increase. By AD 1700 it was back up to the same level as the peak in the early Holocene and has since continued rising. We're now at 1775 parts per billion; nearly three times the low level reached 5,000 years ago. We know that methane levels in the air are driven by natural processes and human behaviour. The rise since AD 1700 has almost entirely been driven by human activity, far exceeding what nature has put into the air: flatulent livestock, rice farming and landfills have all helped drive methane up to today's unprecedented levels. The question is why did the methane levels begin to rise around 5,000 years ago? Some early work had suggested that it was a mixture of changing wetland areas in Siberia and the tropics. This might sound plausible but the devil is in the detail.

The early Holocene peak in methane levels between 11,700 and 8,500 years ago was almost certainly driven by natural processes caused by the changing orbit of the Earth around the Sun. As we'll see in the next chapter, this warming by the Sun drove a strong summer monsoon in the low latitudes. As the monsoon strengthened, basins filled, flooding huge areas to form vast wetlands that produced large amounts of methane. When we look at what methane did further back in the past, something similar seems to have happened in the early part of previous interglacials. Reconstructions of atmospheric methane from gas bubbles trapped in the Antarctic Vostok ice core show an early peak in methane during the first warming of each interglacial. But in contrast to the Holocene, the levels drop away into the next ice age. This all suggests the rise in atmospheric methane 5,000 years ago is a bit odd.

There is also a subtle but important difference between the Holocene methane levels preserved in the Greenland and Antarctic gas bubbles. The short lifespan of methane means it gets broken down before it can be thoroughly mixed throughout the atmosphere. The result is that although the same peaks and troughs are found in different ice cores, the absolute amounts are slightly different. Most of the world's wetlands are in the northern hemisphere; they're therefore closer to Greenland, with the result that there tends to be from 5 to 10% more methane in these gas bubbles than those in Antarctica. Over the past 5,000 years, however, the difference has decreased, suggesting the Siberian wetlands aren't the cause of the rise. This, coupled with the changing orbit of the Earth drying out the tropics, means that a natural change in wetland area can't explain the weird methane increase. The consensus that methane levels after 5,000 years ago were driven by natural changes was finally blown out of the water with a series of papers published by Ruddiman. In 2001, with his undergraduate student Jonathan Thomson, Ruddiman proposed that the cause of the rise 5,000 years ago was human.

Ruddiman and Thomson argued that our ancestors started rice paddy farming on a large scale in southeast Asia 5,000 years ago. When they did this, they grew the rice in flooded fields to protect them from weeds and the increasingly drier conditions. It was a recipe for arresting the natural decline in wetlands and inadvertently pumped more methane into the atmosphere.

Methane was not the only thing that was doing something unexpected in the Holocene. Carbon dioxide also bucked a trend. If we again look at what happened during past interglacials, we find that carbon dioxide levels in the atmosphere also peaked with the insolation from the Sun and then fell away. At the start of the Holocene, the trend in carbon dioxide followed the pattern of previous warm periods, peaking in the early interglacial around 10,500 years ago at 270 parts per million and then falling away. But 8,000 years ago, carbon dioxide levels began rising again. If the previous interglacials are anything to go by, the levels should still be falling. What happened?

The timing of the carbon dioxide increase is suspiciously close to when farming kicked off in earnest across Europe. If we follow Ruddiman's argument, when humans started clearing the land of forests, they released the carbon from the trees and soils and put it back into the air, increasing the carbon dioxide levels in the atmosphere and reversing the natural trend. Ruddiman has suggested that clearing the forests wouldn't have been enough in itself to cause all of the observed change in carbon dioxide, but would have set off a series of natural positive feedbacks that pushed levels higher still.

In 2000, Nobel prize winner Paul Crutzen and Eugene Stoermer argued that we are now in a new geological epoch that they labelled the Anthropocene; a period when human activity had significantly impacted on the world's climate and environment, shifting it out of its natural state. They suggested on the basis of the increasing greenhouse gas levels that a good place to define where it started was AD 1800; this was when greenhouse gas levels started to push beyond what had been seen during glacial–interglacial cycles. But by Ruddiman's argument, there should be a lot less of these greenhouse gases in the atmosphere; they were only at the upper limits of their natural range at the end of the 18th century because our ancestors had been clearing forests and growing rice in paddy fields for millennia. Instead, Ruddiman has made a strong case that the Anthropocene can be pushed back to 8,000 years ago, when carbon dioxide levels started to increase again. It does make you wonder whether we've been playing with fire all along and not realized it.

At the end of the last ice age, the world's climate lurched from one state to another. The start of the Holocene continued to be a bumpy ride. Big changes took place in the early part of the interglacial as the climate and environment moved towards a new equilibrium.

Due to the Earth's varying orbit in the early Holocene, the Arctic received a relatively large amount of heat from the Sun, causing widespread melting of sea ice. The changes we're seeing in this part of the world today, however, can't be blamed on the Earth's orbit; this region is now getting less energy from the Sun than 10,000 years ago. In spite of this, what happened to the Arctic sea ice back in the past does provide an important context for the future. We know that sea ice in the Arctic began to form 45 million years ago. These conditions led to the evolution of a whole suite of mammals that were adapted to life on and around the ice. Posters of cute and cuddly polar bears adorning bedrooms around the world would not have been possible without this momentous shift to sea ice. More important, however, is the fact that we still have a world of polar bears and the less cuddly walruses and bowhead whales, showing that the ice hasn't completely disappeared during the past few million years. Yet this might be about to change.

The Northwest Passage is one of those names that conjure up romantic images of bearded explorers with a passion for cannibalism. The Victorian dream of taking a sailing ship through the Arctic to the North Pacific without all that tedious mucking about with South America took quite a few lives. Even with the construction of the Panama Canal, such a route would shave nearly 4,000 kilometres off a trip between Europe to Asia. In 1906, the Norwegian Roald Amundsen managed to negotiate his way through the icebergs and lived to tell the tale, but it took him two years. Until recently, not enough ice has melted through the summer to make the trip possible for routine shipping. This all changed in 2007 when the European Space Agency announced a navigable route had opened up through the ice for the first time since records began.

The amount of winter and summer sea ice in the Arctic has been on the decrease for the past three decades. Almost all computer models agree that the scale of ice coverage will continue to decline in the 21st century. Although some ice will reform during winter, the Intergovernmental Panel on Climate Change has compiled a suite of computer models that suggest the Arctic seas will essentially become ice-free during the summertime between 2050 and 2100. More recent work by Julienne Stroeve and colleagues at the University of Colorado makes even more sobering reading. They've looked in detail at the loss of sea ice in the Arctic. The results suggests that if we take the observational data over the past three decades, almost 60% of the lost sea ice can be explained by increasing greenhouse gas levels; far higher than anyone has suggested before. Not only this but the models all underestimate how much of the sea ice has been lost. If this is right, it implies the Arctic is a lot more sensitive to greenhouse gas levels than we might think; something the opening of the Northwest Passage in 2007 strongly supports. If so, a summertime ice-free Arctic in the latter half of this century might be a tad optimistic; one recent model has shown climate feedbacks in the Arctic could have the region free of summer ice as early as 2040; it's not inconceivable it might happen sooner still. We're on a fast track to conditions not seen for at least 8,900 years. Yet even with greenhouse gases playing an increasing important role in the Arctic, not everyone has caught on. Perversely, when the news of the newly opened Northwest Passage was made public, some politicians and media reports seemed to argue there was a plus side: vast reserves of previously inaccessible oil and gas will now be available for drilling. What a dreadful idea.

Even years after the ice has gone, its effects can still be felt. Although the British Ice Sheet was a lot smaller than its Scandinavian cousin, most of the ice was centred over the north of the country. The result was that southern England bulged up at this time. When the ice disappeared, Scotland stared bouncing up while southern England began to sink; Scotland continues to rise by some 1.5 millimetres a year while southern England is sinking at around 1 millimetre a year. This might sound rather innocuous, but the effect continues to play out into the future. Southern England is feeling a double whammy: rising sea levels and sinking land.

London is at real risk. The Thames Barrier was built to defend the city against flooding. When it was first used in the 1980s it was only raised once or twice a year. In 2003 it was raised 19 times in January alone. The combined effect of rising sea levels, sinking land and future increases in winter rainfall means that the Thames Barrier (or a future replacement) is going to get used a heck of a lot more. You'd reasonably expect that this sort of trend might cause a pause in the rate of development along the Thames, but the reverse seems to be the case. In 2003, the British Government announced that it wanted tens of thousands of new homes built along the floodplain by 2016, all downstream of the Thames Barrier. Research done by the Environment Agency of England and Wales estimates that these new homes would add £1 billion to any flood damage if the river defences were breached. It doesn't look like the penny has dropped in some quarters.

Importantly, however, it is from the early Holocene that we get the first hint that human activity can have global consequences. Sometime around 8,000 years ago it looks like we might have started to take control, albeit unintentionally. According to Ruddiman's hypothesis, our ancestors' behaviour resulted in an extra 40 parts per million of carbon dioxide and 250 parts per billion methane being put into the air before industrialization. The known heating effect of these extra greenhouse gases would equate to a warming of around 0.8 °C. This doesn't sound much in itself but climate models suggest that without these extra gases, the natural orbital changes would be putting us into another ice age. By the time of industrialization, our ancestors had pushed the greenhouse gas levels to the upper limits of their natural range. We'd not yet taken the reins from nature, but we were well on the way. I guess if we want to look on the bright side we can always say we've stopped an ice age.

Yet changes continued into the Holocene. Things may have warmed up but the climate wasn't as benign as we might think and could still pack a punch.

Chapter 8
RISE AND FALL

History books hark on about earlier ages when ancient civilizations prospered – and occasionally fought – in bright and sunny climes. The birth of civilization is thought to have happened against a balmy backdrop; it didn't arise when the world was in the middle of an ice age. It all seems to point to a nice easy relationship: warm equals good, cold is bad. It's tempting to simplify the whole messy debate about climate change along the same lines. But it's not as simple as that. Even when the world was warm, it wasn't a smooth ride for our ancestors. Climate continued to vary – albeit not as dramatically as before – and these changes wielded a big stick.

We have to be careful here. The idea that the environment – in particular climate – influences the behaviour of a society is known as environmental determinism. Over the long term, changes in climate can make the difference between a rise or fall in civilization. But some supporters of this idea seem to passionately believe that every slight change in climate has a big knock-on effect. Its opponents are equally vehement and argue that the environment has no role to play; if it gets cold, people light a fire; if it gets dry, people draw on a well. You get the idea. In the few times I get invited to a dinner party, it can be good sport to toss the phrase 'environmental determinism' into a conversation with a colleague and see what happens. Either they'll momentarily look stunned, as if you might have escaped from the local lunatic asylum, and then rant and rave about what a load of rubbish the whole idea is, or they'll nervously scan around, and conspiratorially whisper that they think there might be something in it. I'll be up front right now and state that I don't think that climate is omnipotent. Neither is it impotent. Climate – and the associated changes in the environment – has an impact on people. By the same token, people (as we're seeing today) influence climate. You can't separate one from the other. Over the next couple of chapters I'll try to explain why.

A great place to start is in North Africa. During the last ice age, the Sahara was a lot larger and drier than it is today; the desert expanded to cover a third of Africa's surface. Then a big change happened. If we were able to go back 10,000 years we might be surprised at what would greet us. The Saharan region wasn't nearly as dry as today. In fact, it would have been very much wetter. This might seem a bit odd because an arid Sahara seems like one of those unassailable facts; a bit like a wet British summer. So why did the sand dunes fall back and rainfall increase in the early Holocene?

Up till now we've had a good look at climate changes recorded in ice cores from Greenland and Antarctica. Although these are fantastic archives, they are not as useful as we might like for looking at climate change in areas inhabited by early civilizations and societies. In warmer climes, the ice core records can only give us an indirect idea of what happened. We need to look at other natural chronicles.

Fortunately, across North Africa there is a plethora of archives. Depressions in the ground testify to where lakes once existed. Some were small, others humungous. Today's Lake Chad is just a shadow of the MegaChad in the early Holocene period; this enormous lake covered an area of at least 330,000 square kilometers; beaches, spits and deltas litter a region where sand is now king. Ancient sediments contain pollen grains that show much of the Sahara was covered in verdant grasslands and shrubs. And if that wasn't enough, there is ample evidence of people living within a menagerie that included hippos, giraffes and elephants. It all indicated there had been lots of rain. Welcome to the African Humid Period.

We might think that now we're in a warm period we can forget about those irritating changes in the way the Earth orbits the Sun. It's tempting to think that Milankovitch's ideas can be forgotten and we can get on with looking at other causes of climate change. Although it's true that orbital changes are the driver of climate cycles over the long term, things don't just stop when we hit an interglacial. As we saw in the Arctic, changes in tilt, wobble and eccentricity all contribute to changing the way heat from the Sun is distributed on the surface. The result is that different parts of the world can feel warmer at different times. Because of the precession of the equinoxes, the northern hemisphere summer of the early Holocene found itself in that part of the Earth's orbit closest to the Sun. Summer heat gradually increased across North Africa, reaching a high point between 11,000 and 10,000 years ago. At its peak, the amount of insolation from the Sun was 8% greater than today. This might not sound an awful lot, but it had a disproportionate effect.

A clue to the changing climate of the Sahara lies in the fact that much of the rain that falls in the region over a year comes from the monsoon in summer (Figure 8.1). This is a wonderfully simple but effective way of moving lots of water from the oceans onto the land. As the land starts to heat up at the start of the summer, the air also warms up. This has a big knock-on effect: the air expands, becomes less dense and then rises. The consequence of all this is that there is a lot less air pushing down on the African landscape. Air has to come in from somewhere to replace the stuff that's rising. Fresh air sweeps in from offshore. But this air isn't dry; it contains lots of moisture from all of the evaporation that's been going on at the ocean's surface. Suddenly we start to see a cycle develop. The moist ocean air hits Africa, gets heated by the land and then also rises. As it does so, it cools and the evaporated water condenses, forming storm clouds that dump their load over Africa. The rising air eventually cools and sinks over the ocean. The whole thing continues until the Sun loses its bite and the land is no longer

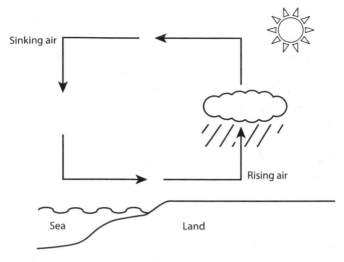

Figure 8.1 The monsoon system.

hotter than the ocean. How far the ocean air can penetrate inland decides how much of the continent receives rain.

In 1981, John Kutzbach at the University of Wisconsin took what happens today and proposed that a stronger system operated in the early Holocene. Because the summer Sun was much stronger at this time, the monsoon was far more intense: the land warmed up a lot more, so more air – and moisture – was bought in from the sea. A warmer land surface meant more air penetrating deeper into the continent, bringing more rain with it. It seemed to explain why the Sahara was green. An 8% increase in summer heat seemed to result in a 40% increase in rainfall. It was an elegant solution and seemed to fit the bill. But there was just one problem. No matter how hard the scientific community tried, they just couldn't get the climate models to explain the lakes seen across the Saharan region. Forty per cent wasn't enough. Changes in the Earth's orbit alone couldn't get enough rain into northern Africa. There had to be feedbacks.

It now looks like other processes exaggerated the effect of the changing orbit. When climate does change it often has a knock-on effect. North Africa at this time was no exception. For a start, more summer heat doesn't just mean that the land is the only place warming up. The ocean would also have got warmer, meaning more evaporation, so more moisture would have been delivered into the Sahara. As high rainfall became the norm, more plants would have grown across the landscape. Soils would have developed, holding onto the moisture for longer and near the surface: rather than draining away as it does in large parts of today's Sahara. It was a recipe for recycling water back into the air and strengthening the monsoon. When all this is added back into the mix, rainfall would have been roughly twice what it is today.

It was in 1833 that a young Charles Darwin realized that the frequent haziness over the eastern tropical Atlantic was caused by dust from Africa. We now know the quantities that can be thrown up into the air are enormous; today some 400 million tonnes of the stuff comes off northwest Africa each year. If the wind is blowing in the right direction it's not uncommon to find a blanket of red dust on the bonnets of cars as far away as the UK. Importantly, the amount of dust varies depending on the size of the desert. To look at how this might have changed in the past, Peter deMenocal and colleagues at the Lamont-Doherty Earth Observatory at Columbia University probed off the west African coast and analyzed an ocean core that's inspiringly known as 658C. The site is perfectly located to work out what happened when. Unlike most ocean cores, the sediments in 658C accumulated at a lightning rate of 18 centimetres every thousand years (that's fast for the seabed) meaning that finely sampled layers could give a detailed record. It also lies right under the main plume of dust. By looking at the changing dust content down through the ocean muds it was possible to work out what was happening on land. The weaker the African summer monsoon, the larger the desert and the more dust that would have hurtled offshore. The results showed that the African monsoon story was far from simple.

You'd be forgiven for thinking that as the summer Sun got stronger, the monsoon would act in proportion: the stronger the Sun's influence, the stronger the monsoon. But this wasn't the case. It looks like the monsoon took a while to respond. As Africa came out of the last ice age, the western Sahara stayed very arid until around 15,000 years ago. Then, when the region got more than about 4% extra summer heat, the monsoon suddenly became disproportionally stronger; there was suddenly a massive collapse in the amount of dust coming offshore. The change was abrupt. Once this threshold was crossed, it looks like everything described above – the warmer sea, the vegetation and the soils – suddenly all worked in unison. The Sahara greened. There was the odd blip during the Younger Dryas but the summer Sun was strong enough to override its effects as soon as the North Atlantic chill had passed. Things then carried on happily for several more thousand years until the summer heat dropped back down to below a level of 4% greater than today. The seas, vegetation and soils all seemed to stop working together around 5,300 years ago. The monsoon became far less effective and dust levels ramped back up to where they had been in the ice age. It all seems to show that the African monsoon works at two different levels. We're in the low gear at the moment. At other times, it can green a desert.

All these changes had a big impact on people living in the region. Across the Sahara there is ample evidence that people existed in the region for thousands of years. There's a fascinating trend in the estimated 10 million paintings and engravings that grace the rocks of the Sahara. A great example is the World Heritage site Tassili n'Ajjer in Algeria which has some 15,000 pieces of art. The drawings and engravings show a clear transition from the early Holocene, when wild animals such as buffalo, giraffes and antelopes were common, to a later period when domesticated livestock like cattle, goats and sheep became the norm. There are even some fabulous scenes showing people dancing to music. It's a remarkable place. But crucially for us, the themes also show a clear progression as the environment responded to the changing monsoon.

In the east Sahara, archaeologists Rudolph Kuper and Stefan Kröplein at the University of Cologne have looked at how people responded to these changes in the monsoon. It's a great story and shows the extent to which people have successfully adapted to a changing climate over time. The results confirm that very few people were visiting the Sahara at the end of the last ice age because of the extremely arid conditions. But things changed abruptly 10,500 years ago when increased monsoonal rainfall caused a greening of the desert in the eastern Sahara. As the buffalo, giraffes and antelopes moved in, prehistoric people rushed in for the kill. It was a golden age of abundant sun, grass and animals. This rich landscape seems to have persevered for a few thousand years, but it all turned sour when the rains started to fail. Between 7,000 and 5,300 years ago, the monsoon became less effective at penetrating the eastern Sahara and the desert began to expand. Those who stayed had to keep mobile. Livestock became a popular way of tracking water in oases and more mountainous areas. There was no point in staying in one place; the environment was too fickle. Most of the desert, however, was simply evacuated and many people migrated south into northern Sudan, following the retreating monsoon. Migration was the key; the mantra was keep with the rains and you should be all right. The people adapted. Most left.

The Nile valley was one of the few refugees left beyond the retreating monsoon; the sudden settlement and development of the pharaohs' civilization along the Nile looks like it only happened when people could no longer survive the full desert conditions around 5,300 years ago. Before then the Nile had been too boggy and forested to prosper in. With the drying that followed, the Nile suddenly became one huge oasis that has continued through to today. Fascinatingly, this leads to an interesting conclusion. Nick Brooks at the University of East Anglia has suggested that a complex society like that of the Egyptians only came about because of climate change. It was when the monsoon started to weaken that people flocked to the Nile. Lots of people had to be organized if the resources were to be effectively used. A political system had to be implemented and a hierarchy established. A benign stable Holocene climate doesn't look like it was the cause of civilization. Civilization, Brooks argues, was an adaptation to increasing aridity.

Something big certainly looks like it happened around 5,300 years ago in North Africa. But is this true of everywhere? What about in the tropics? It's not just an academic question. The tropics contain half of the world's land surface. They're also home to 70% of the world's population. And if that wasn't enough, the tropics are the energy powerhouse of the world. The high latitudes might be crucial for putting the world into and out of an ice age, but it's the tropics that drive the atmospheric circulation of our planet. If we want to learn from the past it's important to see what happened in this part of the world.

Over the course of a year, the tropics receive more energy from the Sun than they lose to space. It's all down to angles. Because of the Earth's angle of tilt, the tropics are almost perpendicular to the Sun all year round. The top of the atmosphere gets the full complement of possible energy from our star with the result that the surface stays relatively warm throughout the year. In contrast, the poles only receive sunlight during the six months around summer; as the Earth rotates through the year, no sunlight falls on the North or South Pole during winter. The bigger the temperature difference between the tropics and the poles, the more heat that has to be transported to high latitudes. Huge amounts of water are evaporated from tropical waters and the energy is temporarily locked up by the moisture within the air and transported polewards. Big storms are an extremely effective way of moving this heat to higher latitudes. But as a hemisphere moves into summertime, the difference in temperature between the tropics and pole decreases. Less energy has to be moved about so the number of storms falls away.

But this isn't the full story. There can be big differences in the amount of rain falling across the tropics, and the 'normal' pattern can change at a moment's notice, influencing the rest of the planet. Probably the single greatest cause of change is El Niño, a phenomenon that disrupts the tropical Pacific once every three to eight years.

El Niño is arguably the best known but most disliked feature of our world's climate system. The name is instantly recognizable; there are few brand names that can compete with its exposure. As a result, the media love it. It's something people both fear and abhor in equal measure. If an unusual climate event has happened, El Niño can be paraded as the cause. Depending on your geographical perspective, it can cause droughts, floods or warming. It's often blamed for some truly cataclysmic events in history; a severe El Niño in the late 18th century was supposedly the cause of French riots that led to revolution; more icebergs in the North Atlantic during 1912 allegedly caused the sinking of the *Titanic*. In climate change, it pretty much catches all but as a result is frequently misunderstood. The name alone is enough to get some people upset, often without them realizing what it is and how it works. In 1998, there was a fantastic report of some poor guy living in Niporno, California, called Al Nino. During a particularly intense El Niño, people had apparently found Al in the telephone book and called to castigate him for the dreadful climate. Sometimes you have to wonder whether everyone's neurons are firing properly.

The name 'El Niño' comes from the Spanish for 'Christ Child' after it was noticed that big changes in the ocean's temperature seem to happen off the west coast of South America around Christmas time. These big changes were first formally described in 1816, when the Governor of St Helena suggested that droughts stretching from India to the Caribbean might have a common cause. But it took another century before anyone really came up with a mechanism to explain how such disparate areas might be linked. In 1923, Sir Gilbert Walker was in charge of the Indian Meteorological Office when he suggested how atmospheric conditions might change across the region.

Walker's insight can best be understood if we imagine looking at the tropical Pacific Ocean from above. Walker argued that changes in atmospheric pressure across the tropical Pacific drove the circulation. It's in equatorial regions that the trade winds of the tropics converge, forming a belt of low pressure known as the Intertropical Convergence Zone. A large part of what drives these winds in the Pacific is rising air in the west and sinking air in the east. It's a bit like a seesaw. When there's low pressure in the west and high pressure in the east, the trade winds blow towards Asia, a phase known as La Niña (Figure 8.2). The rising air over the west Pacific causes huge thunderstorms, resulting in rainfall that can exceed two metres a year, while in the east, the air sinks over the Pacific, off the Peruvian coast. When the difference in pressure across the Pacific weakens, the winds slacken off and can even reverse. Walker wasn't sure what drove these changes but he suspected that the oceans might be the key.

We now know that in a normal or La Niña state, the winds blowing from the east drive the South Equatorial Current along the surface of the tropical Pacific Ocean. Plotting the ocean temperatures shows that the warmest part normally lies in the west and is usually at a higher sea level; the ocean surface in the east is typically 40 centimetres lower than the west. But this water can't keep on flowing to the west indefinitely; there has to be seawater returning east to replace the stuff stacking up in the west. This balancing act is maintained 100 metres under the surface by a large return flow of water towards the west coast of South America known as the Equatorial Undercurrent. The Equatorial Undercurrent picks up nutrients on its travels so that by the time it converges on South America and wells up it can support large populations of fish that are a valuable part of the local economy.

Normal and La Niña

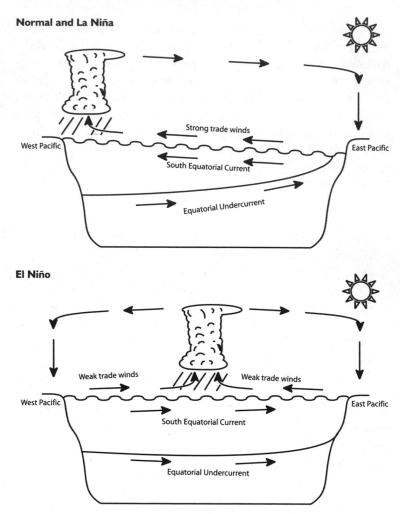

Figure 8.2 The changing states of the Pacific Ocean.

These two ocean currents have distinct differences in temperature. The surface waters are warm; the water at depth is cold. This has important implications, because during an El Niño, the centre of ocean warmth moves out into the central Pacific, driving a cascade of change. Accompanying the weakening of the trade winds, the upwelling of deep nutrient-rich cold water is significantly reduced off the coast of South America, causing a crash in local fish populations. Further afield, the atmosphere gets thoroughly screwed up, causing droughts in the west Pacific, including most of the east coast of Australia, while the southwest USA and northern South America suddenly experience a lot more rainfall.

Bathed in the warm waters of the west Pacific, corals provide a valuable record of what El Niño did in the past. As corals grow, they build up a skeleton of calcium carbonate. And just as we saw with forams earlier in the book, the oxygen needed for the carbonate comes from the water. In the west Pacific region, El Niños cause less rain to fall and a cooling of the ocean. As a result, there tends to be more of the heavy oxygen isotope in the water. When normal conditions prevail, the water is warm and there is lots of rainfall, so more of the lighter oxygen is fixed in the coral skeleton. Changes in the proportion of the different isotopes gives a great handle on whether there were El Niños in the past. The result is that a scientist can come along and do what has to be one of the best jobs in the profession: scuba diving in the tropics to core corals. In the beautiful white bar of rock that's collected is a year-by-year record of what the conditions were like in the ocean when the coral was alive.

In 2001, Sandy Tudhope at the University of Edinburgh and colleagues reported a fantastic study where they had collected a large series of coral cores from across the west Pacific. They measured the oxygen isotopes in the carbonate to develop a record of El Niño for different segments of time stretching back to the beginning of the Eemian. What was immediately clear was that there had been El Niños in the past; it wasn't just something that happened today. More importantly, however, was their observation that the size and frequency of El Niños was markedly higher during the 20th century compared with the past. Follow-up work by other researchers shows that in the early Holocene, El Niños were less frequent and not nearly so intense as today. So what changed and when? Unfortunately, corals aren't our best bet for finding out in the near future. Although they give a fabulous record of the conditions during their life, the largest species can only live for a hundred years or so. Although this is impressive, it's hard to get a full picture of what was happening in the ocean over thousands of years without a lot more samples being collected.

All is not lost, however. We can look elsewhere. Probably one of the most promising records is a rather small lake known as Laguna Pallcacocha in the west Ecuadorian Andes. During an El Niño, this region gets a lot more rainfall. As a result, a lot more sediment gets washed off the mountainsides and dumped into the lake. In 2002, a team led by Chris Moy at Syracuse University and Don Rodbell at Union College in the USA looked at the sediments laid down in the lake over 12,000 years. In it they found a big jump in the number of layers washed off the mountains around 5,000 years ago. Before this time, El Niños seem to have happened on average once very 15 years. Five thousand years ago, however, El Niño started to behave a lot more like today, taking place once every 3 to 8 years. This marked change looks like it had a big impact on people living across the region. On the Peruvian coast, archaeological finds show local fishermen changed their diet of fish and seashells in response to the change in upwelling. In Queensland, Australia, the drier landscape meant people started foraging further afield to get the same sustenance from the land.

Changes in the climate were also happening on top of the mountains. American Lonnie Thompson from Ohio State University, with partner Ellen Mosley-Thompson and a large team of scientists over the years, has pioneered the coring of glaciers and ice caps that perch above verdant forests and deserts of the tropics. It all seems rather incongruous, but with mountains more than 4.5 kilometres high, it can be cold enough for ice to survive through the summer. The tropics might only hold one per cent of all permanent ice, but the information it contains is gold dust.

Unfortunately, these caps are so high, traditional methods of getting to the sites and coring the ice aren't possible. The Quelccaya ice cap in the Peruvian Andes is a classic example. The Andes is home to more than 99% of all tropical ice and Quelccaya is the largest ice cap. The problem is that it sits in the mountains at a height of over 5.5 kilometres above sea level. It's too lofty for a helicopter to transport the coring gear. In the end, Thompson's team had to use pack animals to take the drilling gear up to Quelccaya. But the results have been worth it; an exquisitely detailed record of change in the tropics has been developed for the past 1500 years. Over the years, Thompson has led nearly 50 expeditions into some of the most tortuous and dangerous terrain in the world, collecting records of past climate. Almost every form of transport imaginable has been used to collect and transport this precious material safely to freezers in the lab. Even balloons have been called in to get cores off the mountains; you have to be quick to stop the ice melting once it's been cored.

Through the 1990s, Thompson and his team reported a series of exciting results taken from the ice in the Andes. From Sajama in Bolivia, they showed that at the same time as El Niño was starting to kick in, less snow fell in this part of the Andes; unlike the northern parts, southern Peru and Bolivia get less rainfall when an El Niño strikes. The ice also contained a lot more dust. It all seemed to show that it was getting drier after 5,200 years ago. Meanwhile, on the Peruvian mountain Huascarán, the oxygen isotopes within the ice suggested that it was warmer before 5,200 years ago, falling from a peak in temperatures around 10,000 years ago. Here was evidence that the tropical Andes were picking up the same trends as other parts of the world. A consistent pattern was emerging.

But it's not just the ice cores that give valuable information. The presence of ice alone can give an important insight into the past. Because of its domed shape, Quelccaya is very sensitive to changes in temperature. When temperatures get above freezing the ice cap rapidly retreats. Over the years, Thompson and colleagues have returned to Quelccaya to monitor what's happening. Startlingly, the ice is falling back at the rate of 60 metres a year. As the ice has disappeared, once buried mosses are coming to light. Crucially, these mosses are still rooted to the ground, showing that the ice must have killed them when it surged forward. Thompson's team have radiocarbon dated these remains: the mosses were buried 5,200 years ago. It looks like temperatures had dropped enough at this time for the ice at Quelccaya to grow. Changes were afoot everywhere.

Until quite recently it was common to hear it said that the last 10,000 years or so were a rather boring period to research past climate change. It didn't seem to have the sex and glamour of ice ages. There weren't huge ice sheets swallowing up the landscape, consuming everything in their path. Instead, it just seemed to be warm with a few blips in temperature and rainfall now and again. Looking back, it does seem odd this view persisted for so long. Even as early as the 1960s, it was known that ice around the world had started to readvance during the past 5,000 years or so. The glaciers might not have been enormous, but there was clear evidence they had got bigger over much of this time. A lot of this ice is only just disappearing now.

In the European Alps, the ice has dramatically fallen back, exposing a profusion of archaeological discoveries, including hunting gear, fur and woollen clothing. Many of these finds date back thousands of years. Leather remains of early farmers are so well preserved that they must have been protected from the elements for at least 5,000 years. It looks like we're returning to low levels of ice cover that haven't been seen for thousands of years.

Arguably the most exciting discovery from melting ice was made by two German tourists in September 1991. Much to their surprise these hikers stumbled across the body of a man melting out from the Schnalstal Glacier in the Ötztal Alps on the Austrian–Italian border. The Austrian authorities got in on the act, and after hacking the body out of the ice, took it to the local morgue. The corpse was soon found to be ancient. Radiocarbon dating revealed that the individual had lived sometime between 5,300 and 5,100 years ago; he was not the victim of a recent accident or murder. The new celeb became known as Ötzi after the area in which he had been found. It was clear he'd been a hunter who lived in and around the Alps. It was an archaeologist's dream. Here was a time capsule from prehistory.

It was later realized that Ötzi had been found 90 metres within the Italian side of the border and was duly handed over to the authorities there in 1998. In a final farce, Italian scientists then discovered an arrowhead in Ötzi's shoulder that had been overlooked in Austria despite years of study; he must have been involved in some sort of fight that had ended in his death. Although Ötzi is a fascinating discovery, the critical thing from our point of view is that the body must have been buried very shortly after death. If Ötzi's corpse had lain on the surface for any length of time, it would have rapidly decomposed. Instead, the body is superbly preserved; Ötzi had to have been buried very quickly. This gives us a concrete time frame for the advance of ice and shows that Schnalstal was growing at the same time as Quelccaya and other glaciers around the world.

As the North African monsoon had suddenly weakened with less summer Sun, the ice got a fresh burst of life. Whereas the glaciers and ice sheets had been in retreat since the end of the last ice age, the slightly cooler summers were a lifesaver. It's probable that changes in the Earth's orbit also drove the change in El Niño activity around 5,000 years ago.

But with this long-term shift to more ice, there were also short-term advances and retreats in glaciers that couldn't be explained by changes in the Earth's orbit. Some other control was beating at a faster pace.

These changes weren't just taking place on land. From the late 1990s, Gerard Bond and colleagues of Lamont-Doherty Earth Observatory at Columbia University looked at a number of ocean cores from across the North Atlantic. Even though the ocean had warmed up since the end of the last ice age, Bond and his team showed that every 1500 years or so there'd be a small pulse of rubble to the sea floor; the amount involved, however, was a lot smaller than we saw with the Heinrich events.

It all seems to point to polar waters and icebergs moving south out of the Nordic and Labrador seas. As the bergs melted, the freshwater might have reduced the amount of deepwater being formed in the North Atlantic, while any debris they contained would have been released into the ocean below. Bond suggested that these changes in the North Atlantic were part of the same climate cycle seen during the Heinrich events. The key thing was that the scale of the change was much smaller than before, suggesting that the vast ice sheets had exaggerated the climate response in the past. The question was what was the driver of these small changes?

One possibility is there had been a change in the amount of heat from the Sun. Although the Earth's orbit can affect how much heat we receive from the Sun, it is also possible that our star can vary how much energy it gives off. We know, for instance, that the Sun is now 30% brighter than it was when our planet was formed 4.6 billion years ago. The amount of energy received from the Sun is staggering. The Earth has an estimated recoverable amount of oil totalling 3 trillion barrels; the Sun delivers the equivalent energy in 1.5 days; the amount of energy we use as a species over a year is delivered by the Sun in just one hour. If the amount of energy from the Sun was to change it's not a big leap of faith to imagine it having an impact on the ground. Recent satellite data does show that the amount of energy from the Sun isn't constant. There seems to be a link with the number of spots on the surface of the Sun; every 11 years there is a peak in the number of sunspots and an increase in the amount of energy given off. But the quantities involved are ever so small: the differences of are of the order of 0.1%. They're significant but not huge.

We can test whether these sorts of solar changes were a cause of the trends seen in the North Atlantic. To do this you really need to have records of the climate and the Sun over thousands of years. Ideally both should be precisely and accurately dated so they can be directly compared. How do we get hold of what the Sun was doing? We know Chinese astronomers were counting the number of spots on the Sun's surface nearly 3,000 years ago, but the record isn't complete and doesn't go far enough back in time for our purposes. Fortunately we can use an indirect method to find out what happened.

Earlier in the book I explained how radiocarbon dating works. The radioactive form of carbon is created by high-energy particles from deep space hitting nitrogen gas in our upper atmosphere. The Sun also produces a stream of high-energy particles known as the solar wind. When the Sun is more active, the solar wind gets stronger and sweeps more of these deep space particles away from our planet; when it's really active it can seriously screw up satellite communication and navigation. Crucially for us, the stronger the wind, the fewer the particles that hit the upper atmosphere and the less radiocarbon that's formed. Fortunately, we have a very good idea of how much radiocarbon there has been in the atmosphere and what controls it. By using a computer model of the carbon cycle, it's possible to work out how much the Sun has influenced radiocarbon levels in the atmosphere over thousands of years. We can therefore use changes in atmospheric radiocarbon to back calculate what the Sun was doing in the past.

But we also need a detailed record of what climate was doing at the same time. While at Queen's University Belfast I was extremely lucky to work with a great team headed up by Mike

Baillie. Over 30 years the team had pulled more than 750 oak trees out of bogs across northern Ireland. Over time there seem to have been periods when there were lots of trees, and at other times hardly anything. It looks like the bogs were sometimes dry enough for the trees to grow quite happily on their surface. But at other times, the water tables in the bogs rose, with the result that the population crashed. The trends were identical to those reported from the North Atlantic by Bond. Here was a clear climate signal on the land. The changes in the Irish trees show that when the North Atlantic got colder, westerly winds, with all the rainfall they bring, intensified over Ireland, making the bogs uninhabitable for the trees. We were doubly lucky because the Irish oaks had been used to help reconstruct the amount of radiocarbon in the atmosphere. As a result, we could directly compare what the Sun was doing with the climate changes happening on the ground.

The results were clear: there was no direct response from the Sun. Whatever was causing the changes seen in the North Atlantic, it wasn't obviously due to variations in the Sun's output. There had to be more to it. Hold this thought for the moment, because we'll come back to this in the next chapter.

Equally exciting to me was how these climate cycles might have impacted people living in the region. On the west coast of Ireland, at a small place known as the Céide Fields, are the oldest Neolithic field boundaries in western Europe. Buried below the peat is a huge area marked out by stone walls showing where farming was actively practised around 5,100 years ago. Interestingly, it all seemed to end around 4,500 years ago. The peatlands expanded, taking the stone walls with them. This was a time when the Irish bog oak numbers plummeted. It looks like the wetter conditions might have had an immediate impact on people living in Ireland.

To test whether the climate did have consequences for the people, we had a look at the ages of archaeological sites excavated over the past 50 years. We found there were distinct periods when people were building structures in the landscape. When the climate really dived in the North Atlantic, forts and more permanent settlements were built. It must have been raining almost incessantly and would have been a dreadful time to be a farmer. The simplest explanation is that people responded to the worsening conditions by building structures to protect the few precious resources they had. Some are truly impressive. Dún Aengus is a stone fort that is literally on the edge of a death-defying 100 metre high cliff on the edge of Inishmore, one of the Aran Islands off the west coast of Ireland. You had to defend what little you had with whatever you could. Perhaps people remembered lessons passed on from older generations. Maybe the responses were a natural human reaction to climate change. Either way, it seems likely people quickly got together to defend themselves when times got bad.

The changes seen in the North Atlantic look like they may have had all manner of consequences further afield. In Ireland the cold and wet conditions may have encouraged people to build forts, but elsewhere, different responses were playing out.

The story of Jesus walking on the Sea of Galilee 2,000 years ago is one of the more popular tales in the Bible. But could there be a natural explanation? Doron Nof at Florida State University led a team to look for a physical basis behind Jesus walking on water. They reasoned it was suspicious that the time of this story was during one of the particularly cold periods seen in the North Atlantic. Could it have become cold enough for Jesus to stand on ice? The Sea of Galilee is actually a freshwater lake, and during today's winters it's nigh on impossible to form any ice. The surface waters certainly cool down during wintertime, but the combination of warm temperatures and mixing of the lake prevents any freezing. Every time the surface approaches 4 °C (remember this is when water reaches its maximum density) it sinks and is replaced by warmer water from below. All of the water would have to be at a temperature below 4 °C before the overturning would stop and ice could start to form on the surface. This just can't happen today. But during the cold phases in the North Atlantic it looks like the eastern Mediterranean also cooled down. The drop in temperature, however, wouldn't have been enough in itself to form ice. A possible solution might be found along the western shore. Here salty and warm springs empty into the bottom of the lake. These are so dense that they naturally stay at the bottom of the lake, stopping the surface water from sinking over an area of around 1000 square metres. Nof's team calculated that even though the underlying water may be warm, a cooling of −4 °C for just two days would be enough to form a layer of ice on the surface thick enough to take a person's weight. Any person strolling on an isolated, frozen part of the lake would appear to be walking on water, particularly if it rained after the ice had formed. Could the miracle really have a physical basis? I'll leave that for you to decide, but there's no doubt it's a fascinating idea.

These climate changes also look like they may have played a role in another 'historic' event. Since the 1990s, a team led by Manfred Korfmann has excavated over 100 metres of water tunnels under the ancient Turkish city of Troy; some are as large as 6 metres across. In the tunnels are layers of carbonate that precipitated on the wall surfaces when water flowed over them. By dating the layers it's been possible to establish that the oldest tunnels were at least 4,200 years old, showing that they'd been dug in the earliest stages of Troy's development. But the tunnels weren't always full of water. There were times when no water flowed and these seem to have coincided with periods when the city was abandoned. All in all, it looks like the city went through ten periods of occupation followed by evacuation; when the chips were down, the people left. Interestingly these periods of evacuation were when it was getting colder in the North Atlantic. When it was getting wetter in Ireland it looks like it was getting drier at Troy. Tantalizingly, the fall of Troy described by Homer in *The Illiad* is supposed to have taken place 2,920 years ago. There's no evidence of water flowing in the tunnels at this time and it coincides with a crash in Irish tree numbers. Were the Trojans gasping for water when the Greeks left their wooden horse? Perhaps Helen was the least of the Trojans' problems.

So where does this all leave us? It's clear that even after the last ice age, changes in the Earth's orbit round the Sun continued to play a big role in the climate of the world. In North Africa, warmer summers heated nearby seas, strengthening the monsoon and leading to a greening of the Sahara. These big changes had a huge impact on people living in the region. For several thousand years, good times were to be had. Game was abundant, easy and tax-free. Around 5,300 years ago, however, the summer heat dropped off, the seas cooled and the monsoon suddenly collapsed along with the plant cover. The desert expanded and migration was the key for people on the ground. They had to keep moving, falling back on the Nile or regions south of where the monsoon still penetrated.

In the shorter term there were other climate changes that can't have been driven by a changing orbit. These helped many glaciers round the world advance and then fall back, and Arctic sea ice to move south and then retreat north. In Ireland and the eastern Mediterranean, there were big changes in temperature and rainfall. Relatively small changes had big effects on the ground.

It is possible that the changes we're now causing will lead to another big shift in climate. The ice in the Andes tells us just how precarious a position we're in; these glaciers really are on the edge of existence. The conditions are only just right for ice to exist at all. The result is that the glaciers are often small and respond very quickly to a changing climate. They're the proverbial canaries in a coalmine. Unfortunately, these small glaciers are increasingly becoming shorter, thinner and smaller in volume.

Disappearing glaciers are some of the most powerful images of climate change. Glacier National Park in the United States is fast running out of its namesakes; nearly 80% of its 150 glaciers have gone since 1850. The fact that the Quelccaya ice has retreated to areas where 5,200-year-old mosses are now being uncovered shows that we've now reached temperatures that are without precedent for thousands of years. Quelccaya is not the only low and mid-latitude ice to be suffering. Other work by Thompson and colleagues has shown that Kilimanjaro has had ice on its peak since the end of the last ice age. The gas bubbles in the upper metre of the ice are elongated, as if someone has come along and stretched them out. This can only happen when the surface has melted and then refrozen. There are no other bubbles like this anywhere through the rest of the 12,000 years of ice core. It's unclear whether this warming is directly to blame but more than 82% of the ice area on Kilimanjaro has been lost since 1912. If trends continue, the ice will be gone from this tropical volcano sometime between 2015 and 2020. As glaciers around the world continue to retreat, many people living in mountainous regions will be without a reliable water supply through the year. Nature may have caused their expansion but it's not driving their demise. It looks like we're getting an early warning of the danger we're in.

One point of optimism might be gleaned from plant responses to a carbon-rich atmosphere. With higher carbon dioxide levels, plants won't need to open their stomata as much to fix the same amount of carbon, providing hope that farming may become more productive in the future. Unfortunately, the latest field studies suggest that the benefits might not be as great as first hoped, with results showing crop yields only rising 7–8% with a 50% increase in carbon dioxide; half that originally anticipated. Importantly, however, the reduced number of stomata might mean that plants are better able to deal with drought. If North Africa warms in the future, might we get a stronger monsoon? A greening of the Sahara would potentially act as an important feedback to any strengthening of the monsoon. Perhaps, but it's a close call. Some climate models suggest the region will get wetter, others say drier. Climate modelling led by Martin Claussen at the

Potsdam Institute for Climate Impact Research has suggested that if carbon dioxide levels continue to rise and if the plants aren't disturbed by grazing, nearly half of the Sahara might see some sort of greening. Whether this will happen will depend on how people respond to change. Will the plants be left to grow? Or will the nearest grazer jump on them? It's hard to be optimistic. We haven't got a great track record when it comes to dealing with climate change.

Chapter 9
DROUGHTS, VINES AND FROST FAIRS

It's easy to get hung up on the tag 'global warming'. There's no doubt it's a useful catchphrase for describing the challenges we face, but there's always the risk that our predicament is just seen as warming. Temperature is of course an important facet of the climate, but it's not our only concern. Downpours in the future are likely to vary around the world and throughout the year. The combined effect of changing rainfall and increasing temperature will mean that some regions will get wetter, others drier. Although this might be good news for the bottled water industry it's a worrying trend. So what can the past tell us? Crucially it allows us to understand whether the changes seen today are the result of natural processes or human activity.

A great benefit of looking at the past is that we can also see what effect historical trends in temperature and moisture have had on other cultures and civilizations. A great example is in the southwest of the USA. If you visit the Mesa Verde National Park you'll be greeted with a stunning sight. Built into many of the cliff faces are the ruins of large stone villages, comprising numerous multi-storey dwellings. Arguably the most impressive is the Cliff Palace, a huge building made up of over 150 rooms, incorporating towers, sunken chambers and high ceilings. At another key site, called Chaco Canyon, the scale of construction is vast; amazingly, more than 200,000 conifer trees were used to build the great houses. These are the remains of the ancient Pueblo people called the Anasazi, who flourished in the desert and scrub of the Four Corners region from around AD 850. This is in a region that today only receives around 30 centimetres of rain a year, much of it during the summer monsoon. In such an arid place it's amazing anyone could accomplish so much. Yet for nearly 500 years these great people did just that. By focusing on low-lying floodplains, the Anasazi had a reliable source of water most years, allowing them to grow maize, squash and beans. When the good times boomed, yields were high, the population grew and the culture blossomed.

During the Anasazi's reign in the southwest, however, it wasn't all plain sailing. There were times when the population also crashed. The archaeological evidence points to two key periods where things went badly for the Anasazi. In AD 1150 it looks like large areas were abandoned. In AD 1300 their settlements were finally all abandoned. When the times got bad, it wasn't pretty. At a site called Cowboy Wash in the southwest a grisly discovery was made. In an Anasazi home dating to around AD 1150, seven bodies were discovered dismembered, cooked and apparently eaten. Disconcertingly, the remains were just dumped on the floor afterwards. Later work showed that cannibalism had indeed been practiced: stone tools tested positive for blood while myoglobin (a protein only found in heart and skeleton muscle) was found in a cooking pot and fossilized excrement at the site. It was not a nice way to go. We'll never know precisely why this particular act was carried out, but the fact it happened during a period when the Anasazi were struggling suggests things weren't rosy at the time. Something wasn't right.

To see whether climate may have played a role at this time, we can interrogate trees in the region. By measuring the thickness of the rings across a tree's trunk it's possible to work out what the growing conditions were like for each year: thick rings indicate that it was a great year for growth; a thin or missing ring meant it was a bad growing season and the tree effectively shut down. Because trees within one region experience the same conditions they should show the same pattern of thick and thin rings. By measuring the rings of trees preserved in the landscape it's possible to match up characteristic patterns of growth. It's the ultimate jigsaw puzzle: trees living for several hundred years have one set of rings that can only match with trees growing at the same time. For some trees, a distinctive set of thick and/or thin rings might happen early on in their life; for others it might come near their end. As the patterns overlap, the trees stretch further back in time until a continuous year-by-year record is created, giving an idea of what the climate was like. The potential is enormous. Some trees can live to a staggering age. The bristlecone pine known as 'Methuselah' has lived in the White Mountains of California for over 4,700 years. In Germany, overlapping tree rings of ancient oaks and pines have created a record stretching back 12,410 years.

The first person to suggest all this was a rather eccentric genius called Charles Babbage, 'the father of computing'. In the 19th century, Babbage was leading the charge to build the first counting machine, the forerunner of today's computers. But Babbage was fascinated by almost anything that moved. When not working on his calculating machines he invented a huge array of things including occulting lights in lighthouses and the ophthalmoscope. There was apparently no limit to what Babbage would do for science; he once baked himself in an oven at 129 °C for several minutes just to see what it felt like. Although a great mind, it's fair to say he was not easy to get on with. He was impatient, abhorred street musicians and frequently worried he was being cheated. In 1838, he compiled his *Ninth Bridgewater Treatise* in which was appended a short article on how rings in individual trees might be linked together to make up one long record of climate.

No one really ran with Babbage's ideas until the 20th century, when Andrew Douglass started looking at tree rings in ponderosa pines to see whether changes in the Sun's activity might influence rainfall. Douglass knew of the 11-year cycles in the number of spots on the surface of the Sun and thought this might have an influence on climate. The American Southwest was ideal, Douglass reasoned, because the trees were living at the limits of water availability. Any change in the amount of rainfall would hit their growth and be instantly recognizable in the ring patterns. To help get this information, Douglass started working on the timbers in the Anasazi ruins. After several years he was able to work out when the settlements were built, but equally importantly he recognized periods when trees in the region were struggling to grow. Douglass argued that more than two decades of thin tree rings showed that a major drought had taken place between 1276 and 1299, forcing the Anasazi to abandon the region.

More work has now been done in the region. Ed Cook and colleagues at the Lamont-Doherty Earth Observatory in Columbia University have pulled together a huge tree ring dataset from across the western half of the USA. They've been able to use these results to map which regions experienced drought and when. Interestingly, it was drier than today for most of the time the Anasazi flourished. But Cook's team has also shown that for several decades at a time the region became even more arid (sometimes described as *mega-droughts*), with particularly harsh periods around AD 1150 and 1250. Other researchers in the region have shown similar results. Trees sensitive to the amount of water in the Colorado River have been used to reconstruct changes in past flow; a reconstruction stretching back to AD 762 shows that the 1150s and the end of the 13th century were exceptionally dry. Why is still not entirely certain, but it looks like it might be related to what was happening in the Pacific. Cool, La Niña-like conditions look like they dominated in the eastern Pacific at this time, causing a weakening or failure of the rains over western USA.

It's clear that the Anasazi were capable of dealing with the aridity, but it looks as though they had problems when the conditions went beyond what they were used to. Perhaps the population had boomed during the good times and this had left them particularly vulnerable to any change in rainfall. As the population grew they would have been forced to rely on more marginal areas where water supply was less reliable; when the rains failed for an unusually long length of time, productivity would have crashed. The fallout on the ground must have been dreadful if the find at Cowboy Wash is anything to go by.

But even regions that aren't considered drylands can also suffer drought. Probably one of the best examples of this is from the jungles of central America. In the mid-19th century, American John Stephens and Englishman Frederick Catherwood visited the Yucatan Peninsula and returned with tales of 'lost cities' in the rainforests of Mexico, Guatemala and Honduras. We now know that these ruins were built by the Maya, a remarkably advanced civilization that established large city-states made up of fabulous palaces, plazas and stepped pyramid temples. Even today, the ruins are impressive. In 2007, the northern Yucatan city of Chichen Itza was named one of the new Seven Wonders of the World.

In spite of frequent bloodletting, the Maya flourished in the region during the first millennium AD. The heyday was around AD 750 during what is described as the Classic Maya Period, when the population is estimated to have been somewhere between 3 and 13 million. To support this number of people they had extensive areas of land dedicated to the growing of maize, squash, beans and cacao that needed a reliable supply of water. The cities were built to collect water. In the northern areas, water could also be accessed most of the year from lakes or by digging sinkholes to get at large groundwater supplies. In the lowlands, the Maya were far more dependent on the rain, which was both seasonal and erratic at times. Today the rains fall during the summer when the convergence of trade winds from the two hemispheres – the Intertropical Convergence Zone – moves north and settles over the Yucatan; this meeting of the trade winds results in torrential downpours over the land. But in the winter, this belt of rain moves south and the climate becomes much drier. If one large downpour happens each year, there's no problem. The Maya had a sophisticated network of reservoirs and irrigation channels that could collect and get the water to where it was needed. Settlements could be supported as long as the water supply was dependable. There was only a problem when the rains didn't fall.

Fortunately, we have a good idea of what the Maya were up to. They had a written system based around hieroglyphs; monuments, altars and staircases are covered in the stuff, celebrating successes and tribulations. But at the end of the Classic Period, sometime around AD 900, the monuments and their musings suddenly stopped. The population is estimated to have crashed by 90–99%. The civilization collapsed. As always, a whole host of different ideas have been suggested for what happened. Archaeologists continue to argue over the details. Popular ideas have included a revolt against the ruling elite or invasion from the south. But if so, what was the cause of all the social upheaval?

In 1995, a team led by David Hodell of the University of Florida threw a new idea into the mix. Putting a core through the sediments at the bottom of Lake Chichancanab in the Yucatan Peninsula, they found a distinctive band made up of gypsum. Gypsum is a mineral of calcium sulphate that is common to the region and found dissolved in local river waters flowing into the lake. Most of the time it remains dissolved. But if the lake level drops, the concentration of gypsum increases; if the concentration gets high enough, crystals of gypsum can form, settling on the lake floor. The layer discovered by Hoddell and his colleagues strongly hinted that a major drought had happened; the largest of the past 8,000 years. When this event was dated they found that it took place between AD 800 and 1000; the right time for the end of the Classic Maya.

One of the criticisms against this work was that the dating wasn't that precise; a 200-year drought seemed unlikely. Also, deciphering the hieroglyphs on the Mayan monuments has suggested to some archaeologists that there was a three-stage collapse of the Maya, starting in the lowlands between AD 760 and 810, and culminating in the abandonment of the north around AD 910. A single long drought shouldn't have caused this pattern.

The problem with looking at Lake Chichancanab was that the sediments weren't being laid down fast enough to pick up this sort of detail. What was really needed was a yearly record of rainfall.

Fortunately for our story the crucial detail came just a few years later. Just off the Venezuelan coast is a deep bowl-shaped feature known as the Cariaco Basin. Its offshore position at the northern limit of the Intertropical Convergence Zone means that the Cariaco Basin is a great recorder of what the rain belts have done in this part of the world, including the Yucatan. Every time the summer rains fall, organic matter, silts and clays are washed off the local hills and swept into the sea to form a dark layer that blankets the floor of the Cariaco Basin. During wintertime, the trade winds strengthen and move south, and the rains cease. This has two consequences for the Cariaco. Firstly, there's an abrupt stop to material coming off the land. Secondly, and more crucially, the flow of trade winds over the Cariaco causes an upwelling of deep water, rich in nutrients. Life in the ocean booms, producing lots of silica and carbonate that blanket the floor of the Cariaco with light-coloured sediments. Essentially, black layers mean wet conditions, lighter layers dry. In 2003, Gerald Haug and colleagues at Germany's GeoForschungsZentrum Potsdam reported a study of the Cariaco sediments that had been laid down year by year over the past 2,000 years.

The results from Cariaco showed that the Yucatan was particularly wet between AD 550 and 750. This matches the population explosion known to have happened during the Classic Maya period. While the good times kept coming everything was fine; the abundant water meant that farmers could provide for the much larger population. But when a succession of droughts took place over a century, the productivity of the land must have crashed. The results from Cariaco show that the Intertropical Convergence Zone kept south for several years at a time around AD 760, 810, 860 and 910. These periods would have been particularly dry over the Yucatan as the summer rains failed. Crucially, these droughts appear to have happened close to the time when many of the city-states were abandoned across the Yucatan.

The timing of the reconstructed droughts using Cariaco Basin fits in nicely with the staggered collapse of the Maya. The cities in the centre and south of the Yucatan were most dependent on surface water and were probably abandoned first. The Maya were pushed to the limits of what the land could provide; their population had mushroomed, but the traditional means of collecting water wasn't enough to provide for everyone during a succession of droughts. Social order must have struggled as the rulers could no longer provide the water expected of them. Eventually it seems that even the groundwater supplies in the north were not enough to stave off the collapse of the city-states. The leaders had failed. It was time to leave and it looks like the people did so in droves.

In other regions of the world, it was changes in temperature that caused our ancestors a few headaches. There's a fantastic film called *Erik the Viking* that I often watch with my children. In it, the hero Erik learns that the Sun has been swallowed and the world is in Ragnarök, a time of final battle when almost everyone in the world is fighting. Erik isn't that keen on the rape and pillage. He sets out to wake the gods and end Ragnarök, stop the bloodshed and bring the Sun back to the sky. During this wonderful *Monty Python*esque adventure, Erik does indeed wake the gods, stopping most of the killing and getting the girl. Somewhat surprisingly, this story has a smidgen of climate truth. The Norse Vikings were very aware of the elements and this was woven into a lot of their mythology. Perhaps with good reason; there's a wealth of evidence that climate played a big role in their history.

Here was a European civilization that settled much of the Atlantic seaboard from the late 8th century AD. Although best known for rape and plunder, the Norse were also skilled navigators and farmers. By AD 871 they had settled Iceland and, using this as a springboard, explored the northwest Atlantic. Many of the adventures west of Iceland have come down to us in two Icelandic sagas: The Greenlanders' Saga and Eirik's Saga. The Norse loved their sagas, and in these a hero in the loosest sense of the word is described: Eirik the Red. Unlike his Hollywood counterpart, Eirik was very keen on bloodshed. In fact, the whole reason he left Norway for Iceland was over a small incident of some killings. Once in Iceland, he started to farm, but was soon involved in yet more fighting. He was kicked out of Iceland for three years on pain of death. Eirik had to get out of Iceland and quickly if he wasn't to be executed. He took the only course available to him: he headed west, where tales told of a new land. Eirik needed land and fast, preferably where the law and angry relatives of the deceased couldn't get him. An earlier attempt at settling Greenland by another group had been tried but settling on the inhospitable eastern coast was a total disaster; as with most Norse stories, carnage ensued. Eirik wasn't going to make the same mistake. He spent his three years of exile exploring the west coast of Greenland and found grassy valleys that were ideal for farming. He came up with the name 'Greenland' because he realized 'people would be the more eager to go there if it had a good name' and returned to Iceland with much fanfare. In spite of his homicidal tendencies, he convinced enough of his compatriots that Greenland was the place to be and set off with some 300 would-be colonialists in AD 985. The first fleet of 25 ships certainly had their fair share of problems; only 17 of them made it through the icy northern floes. The majority of those who survived called an area known as the Eastern Settlement home; confusingly this was on the southwest coast. Not everyone put their roots down here though, possibly because of Eirik's known love of violent arguments; several of the ships pushed on up the coast another 640 kilometres and founded the West Settlement.

At first, things seemed to go well. Both settlements prospered. Eirik seemed to calm down and stopped killing people. Word got out that the times were good in Greenland and before long most of the available land was being farmed. At their peak, the Greenland settlements were probably supporting several thousand individuals, farming the land for meat and dairy and exporting walrus ivory, skins and tusks of the narwhal (known as unicorn horns) back to Europe. In 1000, the Vikings on Greenland began converting to Christianity. Before long there was a dedicated bishop and a large building program that culminated in the construction of one cathedral, a monastery and a scattering of churches.

Shortly after Greenland had been settled, a trader called Bjarni Herjolfsson was blown off course from Iceland and in the process appears to have been the first to have sighted North America. He described a heavily wooded land, but intent on reaching Greenland before winter set in, sailed northeast nonstop. Several of his compatriots weren't that impressed with Bjarni's lack of adventure. None more so than Eirik's son, the wonderfully named Leif the Lucky, who on buying Bjarni's boat set off to retrace the previous owner's route. After several days, the sagas describe Leif reaching a land where timber was abundant and the grapes grew wild. The news spread like wildfire; in Greenland these commodities had to be imported from Europe. Leif named the new land Vinland – the Land of Grapes. Future attempts to settle this new land failed, however. The Vikings encountered native Americans who they called skrælings – a politically incorrect word meaning 'wretches'. Although the first attempts to trade were successful, the relationship soon broke down and after three years of fighting the Norse gave up and returned home, apparently never to return. Unfortunately none of the locations described in the sagas have yet been positively identified, but in 1961 the excavation of Viking structures at the Newfoundland site of L'Anse aux Meadows proved once and for all that the Norse had indeed reached North America. They'd beaten Columbus by some 500 years.

But this rash of exploration didn't last. By the late 14th century, European ships were finding it increasingly difficult to reach Greenland through the sea ice that was becoming more common.

As trade links dropped away, the people were hit with a double whammy. With the loosening ties to Europe, agricultural production looks like it began to struggle as well. As they started losing their regular supply of goods from Europe, the archaeological evidence points to a struggling Viking community.

Excavations of farms in the Western Settlement show that the occupants must have experienced the full range of climates Greenland can muster. In the deeper layers of the farm floors have been recovered a species of fly, *Delia fabricii*, which is an indicator of a landscape filled with meadow grass. Times were clearly good at the beginning. But during the later stages of habitation, the insect remains indicate colder conditions. Wood looks like it was in such short supply that it was often reused. Towards the end, conditions appeared to have become so desperate that even the hunting dogs were eaten; the survivors were clearly struggling to find things to eat. When the farms were finally abandoned, the roofs and doors were left to collapse, showing that no one was left to take advantage of the remaining material. In the final upper levels of the excavations, the fly remains switch to those of species living in the cold outdoors. By the middle of the 14th century, the Western Settlement had been lost.

Yet the Greenland Vikings didn't hang on to all their old ways. Although they were Europeans they weren't tied to traditional means of sustenance through thick and thin. In 1999, Jette Arneborg and colleagues from the National Museum of Denmark looked at the remains of 27 Greenland Vikings who had been buried over the 500 years of settlement. By analyzing the carbon isotopic makeup of the bones it's possible to find out where the Vikings were sourcing the protein in their diet from; marine life has a lot more of the heavier carbon isotope than plants and animals living on land. The results from the human bones show a fascinating trend. Over time there was a lot more of the heavier carbon in the bones. Through the period of Greenland occupation, the Viking people changed their diet. Originally, about 20% of their food came from the ocean. Over the ensuing centuries, however, the sea became more important, presumably because the traditional Viking sources of dairy and meat were becoming less reliable. By the end they were getting nearly 80% of their diet from the ocean, most of it probably seals. Unfortunately it doesn't look like it was enough.

Intriguingly, the Vikings in Greenland had company from around AD 1100 when Inuit hunters arrived from Ellesmere Island. The relationship was often fraught and frequently ended in tears. The Inuit were also bestowed with the endearing term skrælings and were not popular with the Vikings, though presumably the feeling was mutual. Importantly, however, whereas the Norse seemed to fail to negotiate the worsening conditions, the Inuit sailed right on through. Why is not entirely clear. It seems likely the Inuit were better adapted to the changing conditions in Greenland. These people subsisted on the land and ice, hunting anything that moved, including sea and land birds, whales, seals and caribou. Compared to the Vikings they were the champions. Whatever the final coup d'état, the worsening conditions look like they were too much for the Norse. By the mid-15th century they'd completely disappeared from Greenland.

But it wasn't just in the extreme northern parts of the Atlantic that changes were afoot. The early period of Viking conquest and settlement seems to have coincided with a time of relatively warm conditions between about AD 900 and 1300. The first person to really get to grips with what was going on was British climatologist Hubert Lamb. In 1965 Lamb drew together historical observations of what the weather was like and suggested that many parts of the world, particularly Europe, experienced a Medieval Warm Epoch or Period. Now it's important to stress here that it wasn't warm all the time. There was still climate variability, with the occasional cold winter thrown in for good measure. The important thing, however, is that it does appear to have been generally warmer.

Lamb and others, for instance, have remarked that it was warm enough for large amounts of wine to be produced in Britain at the time. To have a successful vineyard that produces grapes that don't make battery acid, you need a combination of conditions: the right amount of Sun and rain, and importantly, a mild winter and warm summer. It's not always something we instinctively associate with Britain, yet there is evidence that quaffable wine was being made here in the past. The Romans seem to have had some success fermenting the stuff, but the real boom for the viticulturalists appears to have been between AD 1100 and 1300. At this time there were at least 46 vineyards south of Ely and Gloucester, with a handful further north. It's worth noting that this wasn't the zenith; today there are now 350 vineyards, with the most northerly site being near Durham in West Yorkshire. The mediæval vineyards were, however, successful enough to upset their French competitors who tried to get them abolished in at least one peace treaty.

Fortunately for the French, when the Viking settlements in Greenland began suffering, conditions seemed to have become a little less suitable for the English vineyard. From around 1350 to 1850, it appears to have got a lot wetter and cooler. A large number of British vineyards closed down. In Scandinavia and the European Alps, glaciers started to readvance. In the Arctic, the sea ice apparently came so far south that there are reports of Inuit landing their kayaks on the shores of Scotland. It was a time known as the Little Ice Age.

Just in case of any misunderstanding, it's worth clarifying that although the terms Medieval Warm Period and Little Ice Age are often bandied about, they should really only be used for where they were first defined. Not everywhere was warm or cold at the same time. In New Zealand, it looks like it might have been a bit warmer during the Medieval Warm Period and cooler during the Little Ice Age, but the dates suggest that they were not at exactly the same time as in the North Atlantic. In Antarctica and parts of the tropical west Pacific, it actually seems to have been warmer during the Little Ice Age. These terms have a lot of baggage with them, so you have to be careful when you use them. For ease of discussion I'm going to use them, but strictly in terms of time.

In southern England, the old London Bridge was a medieval contraption that straddled the River Thames and allowed very little flow of water. The result was that when the winters became particularly cold during the Little Ice Age, the Thames could actually freeze over. The opportunity wasn't lost on local entrepreneurs. Frost fairs sprang up, often lasting weeks. You could buy or do almost anything: ice skating, bull baiting, puppet plays, souvenir cards, alcohol. The list was almost endless. There are even reports of a poor elephant being dragged onto the ice during the last frost fair in the winter of 1814. But in 1831 the new London Bridge was opened. The combination of wider arches and warmer winters spelled the end for the frost fairs.

By the 19th century, it was realized that things weren't quite right. In 1834, *The Times* of London published a short article on some work by a researcher called M. Arago at the French Institute who argued that it was relatively cold at the time; to support his case he used the fact that England had once been a known wine producer. He wasn't sure what was driving the change. He considered the Sun but also suggested that more ice from the Arctic might be the cause. Arago was thinking along the right lines.

The Medieval Warm Period and Little Ice Age appear to have been a continuation of the same 1,500 year cycles seen in the North Atlantic during the last ice age and through the Holocene. Not only did these changes directly influence the temperature on the ground, but they also appear to have changed the way the air was circulating, exacerbating the effect. Just as the tropical Pacific is straddled by high- and low-pressure systems, it was realized in the early 20th century that a similar situation existed in the North Atlantic. In the Atlantic, it's changes in the sea level pressure difference between Iceland and the Azores that influence the strength of the westerly winds. This difference is called the North Atlantic Oscillation, or NAO for short. When the pressure difference between the two regions is large, the NAO is defined as being in a positive phase, which results in lots of warm, moist air being bought off the Atlantic onto northern Europe. But when the air pressure systems over Iceland and the Azores are relatively weak, the NAO shifts into a negative phase, and fewer and weaker storms hit northern Europe, cooling the region. A negative NAO seems to have characterized much of the Little Ice Age, particularly between 1650 and 1710 when there were an unusually low number of spots on the Sun – a period known as the Maunder Minimum. At the same time, the Gulf Stream was transporting about 10% less water north. As less heat came up from the tropics, the ice started streaming out of the Arctic. It looks like a whole host of different processes conspired to cool the North Atlantic region.

It's easy to throw your arms about and argue that it got warmer or colder in the past. Many have done so before; some even to dispute that there's anything special about the warming we're seeing today; after all, if the North Atlantic is warming up after the Little Ice Age, the temperatures are going to be higher than they were a few hundred years ago. But the crucial question we need to be asking is whether today's warming is above and beyond what we'd expect. Are the higher temperatures we're seeing today out of the ordinary compared to previous millennia? To do this we need to talk numbers. By how much did the temperature change in the past? It's only by knowing this that we can directly compare what we're observing today with what's gone before. We need to be able to look at years when the climate went to an extreme, and ask whether this is unusual.

To answer these sorts of questions we'd ideally have monthly and yearly records of all the different aspects of our planet's climate stretching back over thousands of years: maximum and minimum temperature, rainfall, wind speed and direction, cloud cover and so on. Unfortunately, meteorology as a science really only got going in the last few hundred years and the network of weather stations we have was only established from the mid-19th century. It's not much of a record to compare today's climate against. As we've seen, all is not lost. Some parts of the natural world have been kindly responding to climate and in some cases, people have kept records of these changes for hundreds of years. They might not have been precise measurements of what the weather was doing but they can be close. The trick is to find those that are documenting events sensitive to climate.

In Europe there's a wealth of information on grape harvest dates. But to get a climate reconstruction from this you need to look at just one type of grape. As any serious wine buff will tell you, different varieties respond to different conditions. In the Burgundy region of France there are yearly records of harvest dates for the Pinot noir grape stretching back to AD 1370. Fortunately for us, the harvest date for Pinot noir is strongly controlled by the spring–summer temperature. In 2004, Isabelle Chuine and colleagues from the Centre National de la Recherche Scientifique at Montpellier took the grape harvest dates to develop a year-by-year record of temperature. Not only did these results confirm that the baking hot European summer of 2003 was nearly 6 °C warmer than normal, but that it was unprecedented in Burgundy over the past 633 years.

Some headway is now being made to develop networks of sites that preserve a record of the different facets of climate in different parts of the world; drought in the western USA is a good example. Ultimately, however, if we want to compare hemispheric trends over the past millennium on a year-by-year basis, temperature is the only game in town. Many biological and physical processes are controlled by temperature, and fortunately some leave a record of this year on year. For instance, in regions where there is plenty of water, the growth of trees is controlled by the temperature of the growing season; in corals, the temperature of the seawater influences the isotopes in their skeletons; and in ice cores, the snow's isotopic makeup is directed by temperature. The list goes on.

The first real attempt to pull all this data together was made by Mike Mann, now at Pennsylvania State University. In the 1990s, he led a team that took the data from northern hemisphere trees, corals and ice cores and calibrated them against temperature. The reconstructions were then combined with historical records and weather station measurements. The result was a yearly record of northern hemisphere surface temperature over the past thousand years. The reconstruction showed that temperatures sloped gently down through most of the past millennium and then rose sharply by 0.7 °C through the 20th century; it wasn't long before the shape of the trend caused it to be dubbed the 'hockey-stick' curve.

Some of the temperature trends that Mann and colleagues found could be explained by changes in heat from the Sun. But this wasn't the only cause. Volcanic eruptions also seemed to play a big role. In 1815, the Indonesian volcano Tambora erupted over several days, spewing vast amounts of sulphur dioxide and ash into the atmosphere. It was the largest eruption in historical times. The world reeled from its impact. The amount of sulphur that was emitted was enough to significantly cool the planet. Unusually cold and wet weather hit many parts of the world the year following Tambora; so much so that 1816 became known as the 'Year without a summer'. In Europe, temperatures dropped by 1–2 °C. In the USA, the weather got so bad that agricultural production crashed; the price of a bushel of wheat went up to $2.50, a value not reached again until 1972. Overall, the temperature of the northern hemisphere dipped by around 0.5 °C. But it wasn't all bad. The grey days of incessant rain inspired some classic pieces of literature. With nothing much to do in the open air, Mary Shelley knocked out *Frankenstein* while Lord Byron composed the poem *Darkness*. It might have been dreary to most, but it was inspirational to some.

Before the work of Mann and his colleagues, it was known that the end of the 20th century was unusually warm. But it wasn't certain just how unusual it was compared with further back in time. The new reconstructions showed that the warming was unprecedented over the past thousand years. The conclusions were dynamite. Changes in the Sun and volcanic eruptions clearly played a role in driving earlier climate changes, but they couldn't explain the big warming trend seen over the past few decades. If these factors had remained dominant, today's temperatures should have been going down. Something had changed. Another factor was driving the warming. The only way to explain it was the increasing amounts of greenhouse gases in the atmosphere. You could have heard a pin drop.

Although the Vikings in Greenland look like they succumbed to colder temperatures, this was unusual. Most cultures and civilizations that have wrestled with shifts in climate look like they were hardest hit by drought. It's the shortage of water that has been the challenge. Since the fall of the Anasazi, the Pacific has experienced more El Niños. The result seems to be that the southwest USA has received a lot more rain. When droughts have taken place in the recent past they've not been nearly as big as those that have gone before. Even the Dust Bowl of the 1930s doesn't rank against the mega-droughts the Anasazi had to contend with. Yet if the Anasazi have a lesson for us, it's if we want to avoid eating our neighbours we'd best be frugal with the water that's available. We're not doing a very good job so far. The reconstruction of river flow through the Colorado shows that the decision on how much water can be pulled from the Great Basin was made during a particularly wet early 20th century. This was such an unusually wet period that it hasn't been seen at any other time over the past 1200 years. This all has worrying implications for the future of a region that's continuing to grow; it shows that the southwest USA can get a lot drier. Climate models are already suggesting this region has started a long-term shift to more arid conditions; it looks like drought is going to be the norm again.

Ultimately, more heat is being trapped in the atmosphere and this is now feeding through to different parts of the world via the ocean, land and ice. It's taking a while for everything to respond.

Much like captaining a big ship, you can give the order to turn, but it's some time before things happen. We're now only just seeing the first change in course. Each year we see this effect on a smaller scale. During a summer solstice, one hemisphere of our planet gets its biggest dose of heat in the year from the Sun. Depending on where you live, it's either on or around 21 June or 21 December. But the hottest day always happens a good month later; it takes some time before this peak in energy from the Sun works its way through the different parts of the world's climate system. The upshot of this is that even if we managed to peg greenhouse gas levels to what they were in 2000, we'd still experience a further global temperature rise of around 0.6 °C because of the time it takes for the oceans to give off the heat they've already absorbed. We've started a big change and it's got a long time to play out.

One suggestion that might buy us some time is to geoengineer a solution by pumping large amounts of sulphate aerosols into the upper atmosphere. It all sounds a bit desperate but in principle it would be like setting off several large volcanic eruptions without all the death and destruction. The aerosols would reflect sunlight back out into space, helping to cool things down until we could cut back our emissions. But unfortunately, just as we saw with Tambora, large eruptions look like they cause major disruptions to the world's water cycle, bringing drought to many parts of the globe. A recent study suggests such a scheme could make things a lot worse. The bottom line is there's no quick fix. There's only one answer: we have to get greenhouse gas levels down and fast.

Chapter 10
THE RISING TIDE

So what does this all mean for the future? In an uncertain world, the past gives us the opportunity to reflect and learn from what has gone before. It's pretty clear we're on thin ice. The air we breathe contains a level of carbon dioxide not seen for at least 650,000 years and may be unprecedented for as long as 3 million years. Our seemingly insatiable need to dump vast amounts of carbon dioxide and other greenhouse gases into the atmosphere is turning our home into a hothouse world. We're pushing our planet out of its comfort zone. If the world was left to its own devices it should now be cooling down. Instead, we've wrested control of the climate and are taking it in the opposite direction. The past tells us of the risks we face. A little more greenhouse gas in the air does not cause a little change in climate. As we've now seen, our planet has one set of feedbacks built on another. A bit of warming can cause a cascade of unintended consequences. With an ever-warming planet, the Earth's ability to soak up greenhouse gases is already lessening, causing yet more warming. If there's one thing we've learnt from the past it's that the world can change at a moment's notice. When we'll reach the tipping point is anyone's guess but we must be getting close.

We're now seeing changes that are without precedent for thousands of years. In 2002, a chunk of Antarctic Peninsula ice known as the Larsen B shelf collapsed into the Weddell Sea, never to be seen again. It might not sound that exciting, but this wasn't just any old piece of ice. This was 3,200 square kilometres, equivalent to the size of Rhode Island. And it wasn't something that happens every Tuesday week. The researchers who went into the area afterwards were able to show that this was a one in 12,000-year event. And it all happened in 31 days; once the shelf started to disintegrate, there was no stopping it. With all we've learnt this still seems incredible. After all the to-ing and fro-ing of temperature over millennia, it was only in the 21st century that the conditions were unusual enough for Larsen B to collapse into the sea.

The Antarctic Peninsula is a fabulous place to observe the changing climate. It's a 1300 kilometre long finger of land that juts far beyond any other part of Antarctica and it's suffering from a severe case of overheating. The region has warmed 2.5 °C since the 1950s; four times the global average. It was all too much for Larsen B. Melt water built up on the surface and squeezed down the crevasses, weakening the ice. The shelf managed to deal with earlier warmings, but this was the final straw. After 12,000 years it was time to say goodbye. It was an extraordinary event in our planet's history.

Larsen B is a strong hint that today's warming is out of the ordinary. But this is just an opening shot. As we've already seen, other dramatic changes are taking place: the Northwest Passage opened up in 2007 for the first time in what looks like 9,000 years; glaciers in the Andes and Europe are falling back to a size not seen for 5,000 years. Meanwhile, temperatures across the northern hemisphere are the highest they've been for at least 1,000 years. These aren't natural changes that can be explained by natural processes. We know from 55 million years ago that putting large quantities of greenhouse gases into the air can lead to catastrophic warming. When all's said and done, we can only explain today's temperatures by taking into account the gases we're pumping into the atmosphere. There's no other way to explain the trends we're seeing. As other parts of the world get warmer, we'll start to see more of these one-in-millennia events.

The Intergovernmental Panel on Climate Change has made a concerted effort to get to grips with what the future will bring. The amount of greenhouse gas levels in the atmosphere will strongly decide the state of the world that our children will inherit and, almost equally importantly, what I'll be able to do in my retirement. If you feel optimistic that emissions will be reined in so that the atmosphere keeps to a mere 600 parts per million carbon dioxide equivalent then this should mean a warming of between 1.1 and 2.9 °C by 2100. But if you fancy we'll just carry on pumping gas into the atmosphere come what may, the levels could hit 1,550 ppm and the temperatures will rise somewhere between 2.4 and 6.4 °C. It's tempting to choose a scenario between these two extremes; at first glance this seems to offer a middle ground that might avoid digging a bunker in the back garden and resorting to popping happy pills. The problem with the mid-range scenario, however, is that the temperatures are still estimated to go up by around 2.5 °C. This might not sound too bad – after all, it's a lot better than 6.4 °C – but look what a rise of this size did to the Antarctic Peninsula. More importantly, this figure of 2.5 °C is just an average. It hides a range of extremes. When the world was around 1 °C warmer during the Eemian, the Arctic sat at 5 °C. Unfortunately, even this mid-range seems unrealistically hopeful. A recent analysis of the latest climate trends shows that carbon dioxide levels, world temperatures and sea level are all rising on the high end of the Intergovernmental Panel on Climate Change projections. If the carbon cycle starts chucking more carbon dioxide up into the air as things warm, we'll face an even greater challenge.

These trends would suggest we're heading for temperatures that were last seen in some parts of the world during the Eemian, but this time it'll be global. If so, the forecast sea level rise of 60 centimetres looks too optimistic. One of the big unknowns is what will happen to the ice sheets in a greenhouse gas-rich world. As we saw earlier, melting shelves in themselves won't cause much of a sea level rise. But when the shelves disappear, the glaciers behind them are no longer held back. When Larsen B collapsed, the glaciers suddenly found they were free to flow as gravity intended and accelerated their flow by three times.

Fortunately, the amount of ice involved was small by world standards so there wasn't a big increase in sea level. But it gives us a taste of the future. The 2100 forecast for sea level is based on the Antarctic and Greenland glaciers not speeding up. But as more shelves melt in other parts of the world, the glaciers must start to accelerate. During the Eemian, the sea levels ended up being 4 to 6 metres higher than today, largely because the polar ice did melt, and some of this must have been from glaciers racing each other to the ocean. How quickly this might happen is still uncertain, but given how fast Larsen B collapsed I'm not going to be buying that beach home I always wanted for my retirement.

Although temperature and its impact on sea level are easy ways to characterize the impact of greenhouse gases, it will be water – or the shortage of it – that will be the greatest challenge. And it will be on the developing world that the hammer will fall hardest. The Intergovernmental Panel on Climate Change predicts that by 2020 – regardless of what we do to greenhouse levels – enough change has been set in motion that between 75 and 220 million people in Africa will suffer from the effects of drought. We're not looking at some abstract time-scale. These changes will affect most of us during our lifetimes. The 20th century is littered with the fallout from major droughts. A recent analysis led by Gemma Narisa of the University of Wisconsin-Madison has shown that at least 30 major droughts of more than 10 years length took place in the last century, virtually all of them in arid and semi-arid areas. As we saw in North Africa 5,000 years ago, once the rains stopped, vegetation and soil cover collapsed, exacerbating the drying effect. It's a feedback we could do without. The Anasazi and Maya show that once a civilization reaches breaking point, it collapses. Cities and towns were abandoned as people went in search of more reliable supplies. These ancient cultures show that civilizations can adapt, but only up to a point. Seventy-five to 220 million Africans have to go somewhere. There's no precedent in the past for the scale of this impending exodus.

But there is one big difference from the past. Although many changes have happened in our planet's history, these were natural. We are now the cause. As such we have a choice: we can change the future. We don't have to go down the road before us. We can still turn back from the gathering storm. The longer we leave cutting emissions, the worse the problem will become. We need to start thinking the unthinkable and change our ways.

It's August 2007 and I'm sitting on a coral beach on the Low Isles. The wind is blowing lightly through the she-oaks that live on this idyllic Great Barrier Reef island. There's only a little bit of shade but the place is filled with people of all ages sitting, walking, and snorkelling. The reef off the island is a cornucopia of shapes, sizes and kaleidoscopic colours. It's a breathtaking place.

As I sit here musing on what I've learnt writing this book, I do wonder what the future might bring. My children, Cara and Robert, are just starting their lives. I'm curious what sort of world their children will inherit. As the sea levels inevitably rise, will my grandchildren be able to enjoy this great icon of the world? In spite of all the sea level and temperature changes that have happened over the last half million years, the Reef has managed to pick itself up and start again. Will it be able to adapt to the cascade of changes that are heading its way? I hope so.

The tide is turning. We can't afford to drift with it.

FURTHER READING

The amount of science being done in the field of climate change is vast. When you consider all the scientific papers, reports, magazine and books that describe what's going on, you have a fantastic recipe for losing the plot. I found the only way forward was to read as much as possible and ask anyone who was willing and able to give me some pointers. What I've done below is compile a list the most relevant publications used during the writing of this book. Wherever possible I've also tried to list one or two more accessible books. I hope these might be of use to those interested in following up some of the themes. For an up-to-date view on climate change and how we can do something about the problem I'd strongly recommend the excellent and accessible blogs www.celsias.com and www.realclimate.org/.

Introduction

Anderson, K. (1999) The weather prophets: science and reputation in Victorian meteorology. *History of Science*, **37**, 179–216.

Burder, G. F. (1863) Admiral Fitzroy and the weather. *The Times*, 21 January, p. 12.

Flannery, T. (2005) *The Weather Makers: The History and Future Impact of Climate Change*. Text Publishing, Melbourne.

Fitzroy, R. (1863) Admiral Fitzroy on the weather. *The Times*, 27 January, p. 6.

Gribbin, J. and Gribbin, M. (2003) *Fitzroy: The Remarkable Story of Darwin's Captain and the Invention of the Weather Forecast*. Review, London.

Henson, R. (2006) *The Rough Guide to Climate Change*. Rough Guides, London.

Manley, G. (1974) Central England temperatures: monthly means 1659 to 1973. *Quarterly Journal of the Royal Meteorological Society*, **100**, 389–405.

Meehl, G. A. *et al.* (2007) Global climate projections. In: *Climate Change 2007: The Physical Science Basis. Contribution of Working Group I to the Fourth Assessment Report of the Intergovernmental Panel on Climate Change* (eds. S. Solomon *et al.*). Cambridge University Press, Cambridge.

Monbiot, G. (2006) *Heat: How to Stop the Planet Burning.* Allen Lane, London.

Reay, D. (2006) *Climate Change Begins at Home: Life on the Two-Way Street of Global Warming.* Macmillan, London.

Chapter 1: Greenhouse

Ångström, K. (1901) Ueber die abhängigkeit der absorption der gase, besonders der kohlensäure, vin der dichte. *Annals of Physics*, **5**, 163–73.

Archer, D. (2007) Methane hydrates and anthropogenic climate change. *Biogeosciences*, **4**, 993–1057.

Arrhenius, S. (1896) On the influence of carbonic acid in the air upon the temperature of the ground. *The London, Edinburgh and Dublin Philosophical Magazine and Journal of Science*, **41**, 237–76.

Bains, S. *et al.* (2000) Termination of global warmth at the Palaeocene/Eocene boundary through productivity feedback. *Nature*, **407**, 171–4.

Bowen, G. J. *et al.* (2006) Eocene hyperthermal event offers insight into greenhouse warming. *Eos*, **87**, 165–9.

Callendar, G. S. (1938) The artificial production of carbon dioxide and its influence on temperature. *Quarterly Journal of the Royal Meteorological Society*, **64**, 223–40.

Crouch, E. M. *et al.* (2003) The *Apectodinium* acme and terrestrial discharge during the Paleocene-Eocene thermal maximum: new palynological, geochemical and calcareous nanoplankton observations at Tawanui, New Zealand. *Palaeogeography, Palaeoclimatology, Palaeoecology*, **194**, 387–403.

Dickens, G. R. (2004) Hydrocarbon-driven warming. *Nature*, **429**, 513–15.

Emanuel, K. (2007) Phaeton's reins: the human hand in climate change. *Boston Review*, January/February.

Howard, J. (2004) 'Physics and fashion': John Tyndall and his audiences in mid-Victorian Britain. *Studies in History and Philosophy of Science*, **35**, 729–58.

Kennett, J. P. and Stott, L. D. (1991) Abrupt deep-sea warming, palae-oceanographic changes and benthic extinctions at the end of the Palaeocene. *Nature*, **353**, 225–9.

Kent, D. V. *et al.* (2003) A case for a comet impact trigger for the Paleocene/Eocene thermal maximum and carbon isotope excursion. *Earth and Planetary Science Letters*, **211**, 13–26.

Kerr, R. A. (2007) Humongous eruptions linked to dramatic environmental changes. *Science*, **316**, 527.

Koch, P. L. *et al.* (1992) Correlation between isotope records in marine and continental carbon reservoirs near the Palaeocene/Eocene boundary. *Nature*, **358**, 319–22.

Kurtz, A. C. *et al.* (2003) Early Cenozoic decoupling of the global carbon and sulfur cycles. *Paleoceanography*, **18**, doi: 10.1029/2003PA000908.

Leifer, I. *et al.* (2006) Natural marine seepage blowout: Contribution to atmospheric methane. *Global Biogeochemical Cycles*, **20**, doi: 10.1029/2005GB002668.

Lowenstein, T. K. and Demicco, R. V. (2006) Elevated Eocene atmospheric CO_2 and its subsequent decline. *Science*, **313**, 1928.

Meinshausen, M. (2006) What does a 2 °C target mean for greenhouse gas concentrations? A brief analysis based on multi-gas emission pathways and several climate sensitivity uncertainty estimates. In: *Avoiding Dangerous Climate Change* (ed. H. J. Schellnhuber), Cambridge University Press, Cambridge.

Moran, K. *et al.* (2006) The Cenozoic palaeoenvironment of the Arctic Ocean. *Nature*, **441**, 601–5.

Mudge, F. B. (1997) The development of the 'greenhouse' theory of global climate change from Victorian times. *Weather*, **52**, 13–17.

Nunes, F. and Norris, R. D. (2006) Abrupt reversal in ocean overturning during the Palaeocene/Eocene warm period. *Nature*, **439**, 60–3.

Storey, M. *et al.* (2007) Paleocene–Eocene Thermal Maximum and the opening of the northeast Atlantic. *Science*, **316**, 587–9.

Svensen, H. *et al.* (2004) Release of methane from a volcanic basin as a mechanism for initial Eocene global warming. *Nature*, **429**, 542–5.

Tyndall, J. (1861) On the absorption and radiation of heat by gases and vapours, and on the physical connexion of radiation, absorption, and conduction. *Philosophical Transactions*, **151**, 1–36.

Wunsch, C. (2002) What is the thermohaline circulation? *Science*, **298**, 1179–81.

Zachos, J. C. (2003) A transient rise in tropical sea surface temperature during the Paleocene–Eocene Thermal Maximum. *Science*, **302**, 1551–4.

Chapter 2: Snowball

Bodiselitsch, B. *et al.* (2005) Estimating duration and intensity of neo-proterozoic snowball glaciations from Ir anomalies. *Science*, **308**, 239–42.

Condon, D. *et al.* (2005) U-Pb ages from the Neoproterozoic Dou-shantuo Formation, China. *Science*, **308**, 95–8.

Darwin, C. (1987) *Charles Darwin's Notebooks, 1836–1844: Geology, Transmutation of Species, Metaphysical Enquiries* (eds. P. H. Barrett *et al.*) Cornell University Press, Ithaca, New York.

Evans, D. A. D. (2006) Proterozoic low orbital obliquity and axial-dipolar geomagnetic field from evaporite palaeolatitudes. *Nature*, **444**, 51–5.

Furnes, H. *et al.* (2007) A vestige of Earth's oldest ophiolite. *Science*, **315**, 1704–7.

Gibbon, E. (1776) *The Decline and Fall of the Roman Empire: Volume 1.* Encyclopedia Britannica, Chicago (1952).

Hards, V. (2005) *Volcanic Contributions to the Global Carbon Cycle.* British Geological Survey, Occasional Publication No. 10, Keyworth, Nottingham.

Harland, W. B. (2007) Origins and assessment of snowball Earth hypotheses. *Geological Magazine*, **144**, 633–42.

Hoffman, P. F. and Schrag, D. P. (2002) The snowball Earth hypothesis: testing the limits of global change. *Terra Nova*, **14**, 129–55.

Hoffman, P. F. *et al.* (1998) A Neoproterozoic Snowball Earth. *Science*, **281**, 1342–6.

Irving, E. (2005) The role of latitude in mobilism debates. *Proceedings of the National Academy of Sciences*, **102**, 1821–8.

Jacka, F. and Jacka, E. (1988) *Mawson's Antarctic Diaries.* Unwin, London.

Kirschvink, J. (1992) Late Proterozoic low-latitude global glaciation: The Snowball Earth. In: *The Proterozoic Biosphere: A Multi-disciplinary Study* (eds. J. W. Schopf and C. Klein), pp. 51–2. Cambridge University Press, Cambridge.

Knoll, A. H. and Walter, M. R. (1992) Latest Proterozoic stratigraphy and Earth history. *Nature*, **356**, 673–8.

Knoll, A. H. *et al.* (2004) A new period for the geologic time scale. *Science*, **305**, 621–2.

Malin, S. and Barraclough, D. (2000) Gilbert's *De Magnete*: An early study of magnetism and electricity. *Eos*, **81**, 233.

Mawson, D. (1949) The Late Precambrian ice-age and glacial record of the Bibliando Dome. *Journal of the Proceedings of the Royal Society of New South Wales*, **82**, 150–74.

Powell, C. M. *et al.* (1993) Paleomagnetic constraints on timing of the Neoproterozoic breakup of Rodinia and the Cambrian formation of Gondwana. *Geology*, **21**, 889–92.

Reusch, H. (1891) Skuringsmerker og morenegrus eftervist i Finnmarken fra en periode meget eldre end 'istiden'. *Norges Geologiske Undersökelse, Årbok*, **1891**, 78–85.

Rice, A. H. N. *et al.* (2003) Three for the Neoproterozoic: Sturtian, Marinoan and Varangerian glaciations. *Geophysical Research Abstracts*, **5**, 11425.

Sonett, C. P. *et al.* (1996) Late Proterozoic and Paleozoic tides, retreat of the Moon, and rotation of the Earth. *Science*, **273**, 100–4.

Sternberg, J. (2006) Preventing another ice age. *Eos*, **48**, 539–42.

Thomson, J. (1871) On the occurrence of pebbles and boulders of granite in schistose rocks of Islay, Scotland. *Report of the British Association for the Advancement of Science, Liverpool*, 88.

Torsvik, T. H. (2003) The Rodinia jigsaw puzzle. *Science*, **300**, 1379–81.

Walker, G. (2003) *Snowball Earth: The Story of a Maverick Scientist and His Theory of the Global Catastrophe that Spawned Life As We Know It*. Bloomsbury, London.

Wegener, A. (1912) Die entstehung der koninente. *Geologische Rundschau*, **3**, 276–92.

Wegener, A. (1924) *The Origin of Continents and Oceans*. Methuen, London.

Williams, G. E. (1998) Precambrian tidal and glacial clastic deposits: implications for Precambrian Earth–Moon dynamics and palaeoclimate. *Sedimentary Geology*, **120**, 55–74.

Williams, G. E. (2000) Geological constraints on the Precambrian history of Earth's rotation and the Moon's orbit. *Reviews of Geophysics*, **38**, 37–60.

Chapter 3: A bit of a chill

Adhémar, J. (1842) *Révolutions de la Mer, Déluges Périodiques*. Paris.

Anderson, J. B. *et al.* (2002) The Antarctic Ice Sheet during the Last Glacial Maximum and its subsequent retreat history: a review. *Quaternary Science Reviews*, **21**, 49–70.

Baker, P. A. (2002) Trans-Atlantic climate connections. *Science*, **296**, 67–8.

Barrows, T. T. *et al.* (2004) Exposure ages for Pleistocene periglacial deposits in Australia. *Quaternary Science Reviews*, **23**, 697–708.

Barrows, T. T. *et al.* (2001) Late Pleistocene glaciation of the Kosciuszko Massif, Snowy Mountains, Australia. *Quaternary Research*, **55**, 179–89.

Barrows, T. T. *et al.* (2002) The timing of the Last Glacial Maximum in Australia. *Quaternary Science Reviews*, **21**, 159–73.

Bennett, M. R. and Glasser, N. F. (1997) *Glacial Geology: Ice Sheets and Landforms*. John Wiley and Sons, Chichester.

Bentley, M. J. (1999) Volume of Antarctic ice at the Last Glacial Maximum, and its impact on global sea level change. *Quaternary Science Reviews*, **18**, 1569–95.

Bowen, D. Q. *et al.* (2002) New data for the Last Glacial Maximum in Great Britain and Ireland. *Quaternary Science Reviews*, **21**, 89–101.

Bromwich, D. H. *et al.* (2004) Polar MM5 simulations of the winter climate of the Laurentide Ice Sheet at the LGM. *Journal of Climate*, **17**, 3415–33.

Chao, B. F. (1996) 'Concrete' testimony to shifting latitude of the tropics. *Eos*, **9**, 9–11.

Claquin, T. *et al.* (2003) Radiative forcing of climate by ice-age atmospheric dust. *Climate Dynamics*, **20**, 193–202.

Climate: Long-Range Investigation, Mapping and Prediction (CLIMAP) Project Members (1981) Seasonal reconstructions of the Earth's surface at the last glacial maximum. *Geological Society of America Map Chart Series MC-36*.

Croll, J. (1864) On the physical cause of the change of climate during geological epochs. *Philosophical Magazine*, **28**, 121–37.

Croll, J. (1875) *Climate and Time in Their Geological Relations. A Theory of Secular Changes of the Earth's Climate*. New York.

Dyke, A. S. *et al.* (2002) The Laurentide and Innuitian ice sheets during the Last Glacial Maximum. *Quaternary Science Reviews*, **21**, 9–31.

Emiliani, C. (1955) Pleistocene temperatures. *Journal of Geology*, **63**, 538–78.

Eyles, N. (2006) The role of meltwater in glacial processes. *Sedimentary Geology*, **190**, 257–68.

Fleming, J. R. (2006) James Croll in context: The encounter between climate dynamics and geology in the second half of the nineteenth century. *History of Geology*, **3**, 43–54.

Hays, J. D. *et al.* (1976) Variations in the Earth's orbit: pacemaker of the ice ages. *Science*, **194**, 1121–32.

Hostetler, S. *et al.* (2006) Sensitivity of Last Glacial Maximum climate to uncertainties in tropical and subtropical ocean temperatures. *Quaternary Science Reviews*, **25**, 1168–85.

Imbrie, J. and Imbrie, K. P. (1979) *Ice Ages: Solving the Mystery*. Macmillan, London.

Lemke, P. *et al.* (2007) Observations: Changes in Snow, Ice and Frozen Ground. In: *Climate Change 2007: The Physical Science Basis*. Contribution of Working Group I to the Fourth Assessment Report of the Intergovernmental Panel on Climate Change (eds. S. Solomon *et al.*). Cambridge University Press, Cambridge.

Lisiecki, L. E. and Raymo, M. E. (2005) A Pliocene–Pleistocene stack of 57 globally distributed benthic δ^{18}O records. *Paleoceanography*, **20**, doi:10.1029/2004PA001071.

Lowe, J. J. and Walker, M. J. C. (1997) *Reconstructing Quaternary Environments*. Longman, Harlow.

Mackintosh, A. *et al.* (2007) Exposure ages from mountain dipsticks in Mac. Robertson Land, East Antarctica, indicate little change in ice-sheet thickness since the Last Glacial Maximum. *Geology*, **35**, 551–4.

Manabe, S. and Broccoli, A. J. (1985) The influence of continental ice sheets on the climate of an ice age. *Journal of Geophysical Research*, **90**, 2167–90.

Meehl, G. A. *et al.* (2007) Global Climate Projections. In: *Climate Change 2007: The Physical Science Basis*. Contribution of Working Group I to the Fourth Assessment Report of the Intergovernmental Panel on Climate Change (eds. S. Solomon *et al.*). Cambridge University Press, Cambridge.

Milankovitch, M. (1920) *Théorie Mathématique des Phénomènes Thermiques produits par la Radiation Solaire*. Gauthier-Villars, Paris.

Mix, A. C. *et al.* (2001) Environmental processes of the ice age: land, oceans, glaciers (EPILOG). *Quaternary Science Reviews*, **20**, 627–57.

Murphy, J. J. (1876) The glacial climate and the polar ice-cap. *Quarterly Journal of the Geological Society*, **32**, 400–6.

Murton, J. B. *et al.* (2006) Bedrock fracture by ice segregation in cold regions. *Science*, **314**, 1127–9.

Nanson, G. C. *et al.* (1995) Aeolian and fluvial evidence of changing climate and wind patterns during the past 100 ka in the western Simpson Desert, Australia. *Palaeogeography, Palaeoclimatology, Palaeoecology*, **113**, 87–102.

Oerlemans, J. (2005) Antarctic ice volume for the last 740 ka calculated with a simple ice sheet model. *Antarctic Science*, **17**, 281–7.

Raymo, M. E. and Ruddiman, W. F. (1992) Tectonic forcing of late Cenozoic climate. *Nature*, **359**, 117–22.

Shackleton, N. J. (1967) Oxygen isotope analyses and Pleistocene temperatures re-assessed. *Nature*, **215**, 15–17.

Shackleton, N. J. and Opdyke, N. D. (1973) Oxygen isotope and palaeomagnetic stratigraphy of equatorial Pacific core V28-238: oxygen isotope temperature and ice volumes on a 10^5 year and 10^6 year scale. *Quaternary Research*, **3**, 39–55.

Shennan, I. *et al.* (2006) Relative sea-level changes, glacial isostatic modelling and ice-sheet reconstructions from the British Isles since the Last Glacial Maximum. *Journal of Quaternary Science*, **21**, 585–99.

Stanley, S. M. (1995) New horizons for paleontology, with two examples: The rise and fall of the Cretaceous Supertethys and the cause of the modern ice age. *Journal of Paleontology*, **69**, 999–1007.

Stendel, M. and Christensen, J. H. (2002) Impact of global warming on permafrost conditions in a coupled GCM. *Geophysical Research Letters*, **29**, doi:10.1029/2001GL014345.

Svendsen, J. I. *et al.* (1999) Maximum extent of the Eurasian ice sheets in the Barents and Kara Sea region during the Weichselian. *Boreas*, **28**, 234–42.

Turney, C. S. M. *et al.* (2006) Integration of ice core, marine and terrestrial records for the Australian Last Glacial Maximum and Termination: A contribution from the Oz INTIMATE group. *Journal of Quaternary Science*, **21**, 75–61.

Walter, K. M. *et al.* (2006) Methane bubbling from Siberian thaw lakes as a positive feedback to climate warming. *Nature*, **443**, 71–5.

Wang, X. *et al.* (2004) Wet periods in northeastern Brazil over the past 210 kyr linked to distant climate anomalies. *Nature*, **432**, 740–3.

Zhang, R. (2006) How cold were the tropics and subtropics at the Last Glacial Maximum? *Quaternary Science Reviews*, **25**, 1150–1.

Zimov, S. A. *et al.* (2006) Permafrost and the global carbon budget. *Science*, **312**, 1612–13.

Chapter 4: A previous warmth

Anthoff, D. *et al.* (2006) *Global and Regional Exposure to Large Rises in Sea-Level: A Sensitivity Analysis*. Tyndall Centre for Climate Change Research, Working Paper 96.

Barnola, J. M. *et al.* (1987) Vostok ice core provides 160,000-year record of atmospheric CO_2. *Nature*, **329**, 408–14.

Bory, A. J.-M. *et al.* (2002) Seasonal variability in the origin of recent atmospheric mineral dust at NorthGRIP, Greenland. *Earth and Planetary Science Letters*, **196**, 123–34.

Brigham-Grette, J. and Hopkins, D. M. (1995) Emergent marine record and paleoclimate of the Last Interglaciation along the northwestern Alaskan coast. *Quaternary Research*, **43**, 159–73.

Brown, G. I. (1999) *Count Rumford: The Extraordinary Life of a Scientific Genius*. Alan Sutton, Stroud.

Canadell, J. G. *et al.* (2007) Contributions to accelerating atmospheric CO_2 growth from economic activity, carbon intensity, and efficiency of natural sinks. *Proceedings of the National Academy of Sciences*, doi: 10.1073/pnas.0702737104.

Cassar, N. *et al.* (2007) The Southern Ocean biological response to aeolian iron deposition. *Science*, **317**, 1067–70.

Charlson, R. *et al.* (1987) Oceanic phytoplankton, atmospheric sulphur, cloud albedo and climate. *Nature*, **326**, 655–61.

Cox, P. M. *et al.* (2000) Acceleration of global warming due to carbon-cycle feedbacks in a coupled climate model. *Nature*, **408**, 184–7.

Dansgaard, W. (2004) *Frozen Annals: Greenland Ice Sheet Research*. Narayana Press, Odder, Denmark.

Dansgaard, W. (1953) The abundance of ^{18}O in atmospheric water and water vapour. *Tellus*, **5**, 461–9.

Duxbury, N. S. *et al.* (2006) Time machine: Ancient life on Earth and in the Cosmos. *Eos*, **87**, 401–6.

Edwards, M. E. *et al.* (2003) Interglacial extension of the boreal forest limit in the Noatak Valley, northwest Alaska: Evidence from an exhumed river-cut bluff and debris apron. *Arctic, Antarctic and Alpine Research*, **35**, 460–8.

EPICA Community Members (2004) Eight glacial cycles from an Antarctic ice core. *Nature*, **429**, 623–8.

Foster, G. L. and Vance, D. (2006) Negligible glacial-interglacial variation in continental chemical weathering rates. *Nature*, **444**, 918–21.

Friedlingstein, P. *et al.* (2006) Climate-carbon cycle feedback analysis: Results from the C^4MIP model intercomparison. *Journal of Climate*, **19**, 3337–53.

Gascoyne, M. *et al.* (1981) Ipswichian fauna of Victoria Cave and the marine palaeoclimatic record. *Nature*, **294**, 652–4.

Geikie, J. (1874) *The Great Ice Age*. W. Isbister, London.

Geikie, J. (1868) Note on the discovery of *Bos primigenius* in the Lower Boulder-Clay of Scotland. *Geological Magazine*, **5**, 393–4.

Gregory, J. M. *et al.* (2004) Threatened loss of the Greenland ice-sheet. *Nature*, **428**, 616.

Hansen, J. E. (2005) A slippery slope: How much global warming constitutes 'dangerous anthropogenic interference'? *Climatic Change*, **68**, 269–79.

Hansen, J. *et al.* (2006) Global temperature change. *Proceedings of the National Academy of Sciences*, **103**, 14288–93.

Hansen, J. *et al.* (2007) Climate change and trace gases. *Philosophical Transactions of the Royal Society of London*, **365A**, 1925–54.

Harkness, D. D. *et al.* (1977) Radiocarbon dating versus the Leeds Hippopotamus – a cautionary tale. *Proceedings of the Yorkshire Geological Society*, **41**, 223–30.

Harting, P. (1874) De bodem van het Eemdal. *Verslagen en Verhandelingen Koninklijke Academie van Wetenschappen*, **8**, 282–90.

Komar, P. D. and Allan, J. C. (2007) Higher waves along U.S. east coast linked to hurricanes. *Eos*, **88**, 301.

Kukla, G. J. *et al.* (2002) Last interglacial climates. *Quaternary Research*, **58**, 2–13.

Le Quére, C. *et al.* (2007) Saturation of the Southern Ocean CO_2 sink due to recent climate change. *Science*, **316**, 1735–8.

Lea, D. W. *et al.* (2006) Paleoclimate history of Galápagos surface waters over the last 135,000 yr. *Quaternary Science Reviews*, **25**, 1152–67.

Legrand, M. *et al.* (1991) Ice-core record of oceanic emissions of dimethylsulphide during the last climate cycle. *Nature*, **350**, 144–6.

Lovelock, J. (2006) *The Revenge of Gaia*. Allen Lane, London.

Lozhkin, A. V. and Anderson, P. M. (1995) The last interglaciation in northeast Siberia. *Quaternary Research*, **43**, 147–58.

Luthcke, S. B. *et al.* (2006) Recent Greenland ice mass loss by drainage system from satellite gravity observations. *Science*, **314**, 1286–9.

Mercer, J. H. (1978) West Antarctic ice sheet and CO_2 greenhouse effect: a threat of disaster. *Nature*, **271**, 321–5.

Meskhidze, N. and Nenes, A. (2006) Phytoplankton and cloudiness in the Southern Ocean. *Science*, **314**, 1419–23.

Muhs, D. R. *et al.* (2002) Timing and warmth of the Last Interglacial period: new U series evidence from Hawaii and Bermuda and a new fossil compilation for North America. *Quaternary Science Reviews*, **21**, 1355–83.

North Greenland Ice Core Project Members (2004) High-resolution record of Northern Hemisphere climate extending into the last interglacial period. *Nature*, **431**, 147–51.

Otto-Bliesner, B. L. *et al*. (2006) Simulating Arctic climate warmth and icefield retreat in the Interglaciation. *Science*, **311**, 1751–3.

Overpeck, J. T. *et al*. (2006) Paleoclimatic evidence for future ice-sheet instability and sea-level rise. *Science*, **311**, 1747–50.

Petit, J. R. *et al*. (1999) Climate and atmospheric history of the past 420,000 years from the Vostok ice core, Antarctica. *Nature*, **399**, 429–36.

Rahmstorf, S. (2007) A semi-empirical approach to projecting future sea-level rise. *Science*, **315**, 368–70.

Rumford, C. (1804) An account of a curious phenomenon observed on the glaciers of Chamouny; together with some occasional observations concerning propagation of heat in fluids. *Philosophical Transactions of the Royal Society of London*, **94**, 23–9.

Scambos, T. A. *et al*. (2000) The link between climate warming and break-up of ice shelves in the Antarctic Peninsula. *Journal of Glaciology*, **46**, 516–30.

Shackleton, N. J. (1969) The last interglacial in the marine and terrestrial records. *Proceedings of the Royal Society of London*, **174**, 135–54.

Shepherd, A. and Wingham, D. (2007) Recent sea-level contributions of the Antarctic and Greenland ice sheets. *Science*, **315**, 1529–32.

Siegenthaler, U. *et al*. (2005) Stable carbon cycle–climate relationship during the late Pleistocene. *Science*, **310**, 1313–17.

Slott, J. M. *et al*. (2006) Coastline responses to changing storm patterns. *Geophysical Research Letters*, **33**, doi: 10.1029/2006GL027445.

Stainforth, D. A. *et al*. (2005) Uncertainty in predictions of the climate response to rising levels of greenhouse gases. *Nature*, **433**, 403–6.

Torn, M. and Harte, J. (2006) Missing feedbacks, asymmetric uncertainties, and the underestimation of future warming. *Geophysical Research Letters*, **33**, doi: 10.1029/2005GL025540.

Willerslev, E. *et al*. (2007) Ancient biomolecules from deep ice cores reveal a forested southern Greenland. *Science*, **317**, 111–14.

Chapter 5: Atlantic armadas

Alley, R. B. (2007) Wally was right: Predictive ability of the North Atlantic 'Conveyor Belt' hypothesis for abrupt climate change. *Annual Review of Earth and Planetary Sciences*, **35**, 241–72.

Alley, R. B. and MacAyeal, D. R. (1994) Ice-rafted debris associated with binge/purge oscillations of the Laurentide Ice Sheet. *Paleoceanography*, **9**, 503–12.

Andrews, J. T. and Tedesco, K. (1992) Detrital carbonate-rich sediments, northwestern Labrador Sea: Implications for ice-sheet dynamics and iceberg rafting (Heinrich) events in the North Atlantic. *Geology*, **20**, 1087–90.

Barrows, T. T. *et al.* (2007) Long-term sea surface temperature and climate change in the Australian–New Zealand region. *Paleoceanography*, **22**, doi:10.1029/2006PA001328.

Bond, G. *et al.* (1993) Correlations between climate records from north Atlantic sediments and Greenland ice. *Nature*, **365**, 143–7.

Bond, G. *et al.* (1992) Evidence for massive discharges of icebergs into the North Atlantic ocean during the last glacial period. *Nature*, **360**, 245–9.

Bond, G. C. and Lotti, R. (1995) Iceberg discharges into the North Atlantic on millennial time scales during the last glaciation. *Science*, **267**, 1005–10.

Broecker, W. (1987) Unpleasant surprises in the greenhouse? *Nature*, **328**, 123.

Broecker, W. S. (1997) Thermohaline circulation, the Achilles Heel of our climate system: will man-made CO_2 upset the current balance? *Science*, **278**, 1582–8.

Broecker, W. S. (1994) Massive iceberg discharges as triggers for global climate change. *Nature*, **372**, 421–4.

Calov, R. *et al.* (2002) Large-scale instabilities of the Laurentide ice sheet simulated in a fully coupled climate-system model. *Geophysical Research Letters*, **29**, doi:10.1029/2002GL016078.

Chappell, J. (1980) Coral morphology, diversity and reef growth. *Nature*, **286**, 249–52.

Chappell, J. (2002) Sea level changes forced ice breakout in the Last Glacial cycle: new results from coral terraces. *Quaternary Science Reviews*, **21**, 1229–40.

de Abreu, L. *et al.* (2003) Millennial-scale oceanic climate variability off the Western Iberian margin during the last two glacial periods. *Marine Geology*, **196**, 1–20.

Donnelly, J. P. and Woodruff, J. D. (2007) Intense hurricane activity over the past 5,000 years controlled by El Niño and the West African monsoon. *Nature*, **447**, 465–8.

Dowdeswell, J. A. *et al.* (1995) Iceberg production, debris rafting, and the extent and thickness of Heinrich layers (H-1, H-2) in North Atlantic sediments. *Geology*, **23**, 301–4.

Elliot, M. *et al.* (2001) Coherent patterns of ice-rafted debris in the Nordic regions during the last glacial (10–60 ka). *Earth and Planetary Science Letters*, **194**, 151–63.

EPICA Community Members (2004) Eight glacial cycles from an Antarctic ice core. *Nature*, **429**, 623–8.

Frappier, A. B. *et al.* (2007) Stalagmite stable isotope record of recent tropical cyclone events. *Geology*, **35**, 111–14.

Gregory, J. M. *et al.* (2005) A model intercomparison of changes in the Atlantic thermohaline circulation in response to increasing atmospheric CO_2 concentration. *Geophysical Research Letters*, **32**, doi:10.1029/2005GL023209.

Grousset, F. E. *et al.* (2000) Were the North Atlantic Heinrich events triggered by the behavior of the European ice sheets? *Geology*, **28**, 123–6.

Hayne, M. and Chappell, J. (2001) Cyclone frequency during the last 5000 yrs from Curacoa Island Queensland. *Palaeogeography, Palaeoclimatology, Palaeoecology*, **168**, 201–19.

Heinrich, H. (1988) Origin and consequence of cyclic rafting in the northeast Atlantic Ocean during the past 130,000 years. *Quaternary Research*, **29**, 142–52.

Hemming, S. R. (2004) Heinrich events: Massive late Pleistocene detritus layers of the North Atlantic and their global climate imprint. *Reviews of Geophysics*, **42**, doi:10.1029/2003RG000128.

Kanfoush, S. L. *et al.* (2000) Millennial-scale instability of the Antarctic Ice Sheet during the last glaciation. *Science*, **288**, 1815–18.

Kanfoush, S. L. *et al.* (2002) Comparison of ice-rafted debris and physical properties in ODP Site 1094 (South Atlantic) with the Vostok ice core over the last four climatic cycles. *Palaeogeography, Palaeoclimatology, Palaeoecology*, **182**, 329–49.

Knutson, T. R. and Tuleya, R. E. (2004) Impact of CO_2-induced warming on simulated hurricane intensity and precipitation: sensitivity to the choice of climate model and convective parameterization. *Journal of Climate*, **17**, 3477–95.

Knutti, R. *et al.* (2004) Strong hemispheric coupling of glacial climate through freshwater discharge and ocean circulation. *Nature*, **430**, 851–6.

MacAyeal, D. R. (1993) Binge/purge oscillations of the Laurentide ice sheet as a cause of the North Atlantic's Heinrich events. *Paleoceanography*, **8**, 775, 784.

Nott, J. *et al.* (2007) Greater frequency variability of landfalling tropical cyclones at centennial compared to seasonal and decadal scales. *Earth and Planetary Science Letters*, **255**, 367–72.

Rohling, E. J. *et al.* (2004) Similar meltwater contributions to glacial sea level changes from Antarctic and northern ice sheets. *Nature*, **430**, 1016–21.

Seager, R. *et al.* (2002) Is the Gulf Stream responsible for Europe's mild winters? *Quarterly Journal of the Royal Meteorological Society*, **128**, 2563–86.

Siddall, M. *et al.* (2003) Sea-level fluctuations during the last glacial cycle. *Nature*, **423**, 853–8.

Sirocko, F. (2003) Ups and downs in the Red Sea. *Nature*, **423**, 813–14.

Stocker, T. F. and Johnsen, S. J. (2003) A minimum thermodynamic model for the bipolar seesaw. *Paleoceanography*, **18**, doi: 10.1029/2003PA000920.

Stommel, H. (1961) Thermohaline convection with two stable regimes of flow. *Tellus*, **13**, 224–30.

Van Kreveld, S. *et al.* (2000) Potential links between surging ice sheets, circulation changes, and the Dansgaard–Oeschger cycles in the Iminger Sea, 60-18 kyr. *Paleoceanography*, **15**, 425–44.

Visbeck, M. (2007) Power of pull. *Nature*, **447**, 383.

Wunsch, C. (2002) What is the thermohaline circulation? *Science*, **298**, 1179–81.

Chapter 6: A belch and a blast

Anderson, B. and Mackintosh, A. (2006) Temperature change is the major driver of late-glacial and Holocene glacier fluctuations in New Zealand. *Geology*, **34**, 121–4.

Atkinson, T. C. *et al.* (1987) Seasonal temperatures in Britain during the past 22,000 years reconstructed using beetle remains. *Nature*, **352**, 587–92.

Bird, M. I. *et al.* (2005) Palaeoenvironments of insular Southeast Asia during the Last Glacial Period: A savanna corridor in Sundaland? *Quaternary Science Reviews*, **24**, 2228–42.

Blunier, T. *et al.* (1998) Asynchrony of Antarctic and Greenland climate change during the last glacial period. *Nature*, **384**, 739–43.

Broecker, W. S. *et al.* (1989) Routing of meltwater from the Laurentide Ice Sheet during the Younger Dryas cold episode. *Nature*, **341**, 318–21.

Calvo, E. *et al.* (2007) Antarctic deglacial pattern in a 30 kyr record of sea surface temperature offshore South Australia. *Geophysical Research Letters*, **34**, doi:10.1029/2007GL029937.

212 ICE, MUD AND BLOOD

Carpenter, C. P. and Woodcock, M. P. (1981) A detailed investigation of a pingo remnant in western Surrey. *Quaternary Studies*, **1**, 1–26.

Chappellaz, J. *et al.* (1993) Synchronous changes in atmospheric CH_4 and Greenland climate between 40 and 8 kyr BP. *Nature*, **366**, 443–5.

Coope, G. R. (1969) Interprétations climatiques des Coléoptères 'Late Weichselian' dans les Iles Britanniques. *Abstracts, VIII Congress INQUA*, Paris, 146.

Coope, G. R. (1977) Fossil coleopteran assemblages as sensitive indicators of climatic changes during the Devensian (Last) cold stage. *Philosophical Transactions of the Royal Society of London*, **B280**, 313–40.

Coope, G. R. *et al.* (1998) Temperature gradients in northern Europe during the last glacial-Holocene transition (14–9 ^{14}C kyr BP) interpreted from coleopteran assemblages. *Journal of Quaternary Science*, **13**, 419–33.

Dansgaard, W. *et al.* (1969) One thousand centuries of climatic record from Camp Century on the Greenland ice sheet. *Science*, **166**, 377–81.

Denton, G. H. and Hendy, C. H. (1994) Younger Dryas age advance of Franz Josef Glacier in the Southern Alps of New Zealand. *Science*, **264**, 1434–7.

Epstein, S. *et al.* (1970) Antarctic Ice Sheet: Stable isotope analyses of Byrd Station cores and interhemispheric climatic implications. *Science*, **168**, 1570–2.

Fairbanks, R. G. (1989) A 17,000-year glacio-eustatic sea level record: influence of glacial melting rates on the Younger Dryas event and deep-ocean circulation. *Nature*, **342**, 637–42.

Firestone, R. B. *et al.* (2007) Evidence for an extraterrestrial impact 12,900 years ago that contributed to the megafaunal extinctions and the Younger Dryas cooling. *Proceedings of the National Academy of Sciences*, **104**, 16016–21.

Fisher, T. G. and Lowell, T. V. (2006) Questioning the age of the Moorhead Phase in the glacial Lake Agassiz basin. *Quaternary Science Reviews*, **25**, 2688–91.

Hanebuth, T. *et al.* (2000) Rapid flooding of the Sunda Shelf: a Late-Glacial sea-level record. *Science*, **288**, 1033–5.

Haynes Jr, C. V. *et al.* (1999) A Clovis well at the type site 11,500 B.C.: the oldest prehistoric well in America. *Geoarchaeology*, **14**, 455–70.

Hughen, K.A. *et al.* (1996) Rapid climate changes in the tropical Atlantic region during the last deglaciation. *Nature*, **380**, 51–4.

FURTHER READING 213

Isarin, R. F. B. (1997) Permafrost distribution and temperatures in Europe during the Younger Dryas. *Permafrost and Periglacial Processes*, **8**, 313–33.

Jessen, K. (1938) Some west Baltic pollen diagrams. *Quartar*, **1**, 124–39.

Johnsen, S. J. *et al.* (1972) Oxygen isotope profiles through the Antarctic and Greenland ice sheets. *Nature*, **235**, 429–34.

Kennett, J. P. and Shackleton, N. (1975) Laurentide ice sheet meltwater recorded in Gulf of Mexico deep-sea cores. *Science*, **188**, 147–50.

Leverington, D. W. *et al.* (2000) Changes in the bathymetry and volume of glacial Lake Agassiz between 11,000 and 9,300 14 yr B.P. *Quaternary Research*, **54**, 174–81.

Lowe, J. J. *et al.* (1999) The chronology of palaeoenvironmental changes during the Last Glacial-Holocene transition: towards an event stratigraphy for the British Isles. *Journal of the Geological Society of London*, **156**, 397–410.

Lowe, J. J. *et al.* (1995) Direct comparison of UK temperatures and Greenland snow accumulation rates, 15 000–12 000 yr ago. *Journal of Quaternary Science*, **10**, 175–80.

Lowell, T. *et al.* (2005) Testing the Lake Agassiz meltwater trigger for the Younger Dryas. *Eos*, **86**, 365–72.

Mangerud, J. and Landvik, J. Y. (2007) Younger Dryas cirque glaciers in western Spitsbergen: smaller than during the Little Ice Age. *Boreas*, **36**, 278–85.

Mangerud, J. *et al.* (1974) Quaternary stratigraphy of Norden, a proposal for terminology and classification. *Boreas*, **3**, 109–28.

Marchitto, T. M. *et al.* (2007) Marine radiocarbon evidence for the mechanism of deglacial atmospheric CO_2 rise. *Science*, **316**, 1456–9.

Mercer, J. H. (1969) The Allerød Oscillation: a European climatic anomaly? *Arctic and Alpine Research*, **1**, 227–34.

Moalem, S. *et al.* (2005) The sweet thing about Type 1 diabetes: a cryoprotective evolutionary adaptation. *Medical Hypotheses*, **65**, 8–16.

Peltier, W. R. *et al.* (2006) Atlantic meridional overturning and climate response to Arctic Ocean freshening. *Geophysical Research Letters*, **33**, doi: 10.1029/2005GL025251.

Rasmussen, S. O. *et al.* (2006) A new Greenland ice core chronology for the last glacial termination. *Journal of Geophysical Research*, **111**, doi: 10.1029/2005JD006079.

Rind, D. (1998) Latitudinal temperature gradients and climate change. *Journal of Geophysical Research*, **103**, 5943–71.

Rooth, C. (1982) Hydrology and ocean circulation. *Progress in Oceanography*, **7**, 131–49.

Ruddiman, W. F. and McIntyre, A. (1981) The North Atlantic during the last deglaciation. *Palaeogeography, Palaeoclimatology, Palaeoecology*, **35**, 145–214.

Sissons, J. B. (1979) Palaeoclimatic inferences from former glaciers in Scotland and the Lake District. *Nature*, **278**, 518–21.

Stanford, J. D. *et al.* (2006) Timing of meltwater pulse 1a and climate responses to meltwater injections. *Paleoceanography*, **21**, doi: 10.1029/2006PA001340.

Stott, L. *et al.* (2007) Southern hemisphere and deep-sea warming led deglacial atmospheric CO_2 rise and tropical warming. *Science*, **318**, 435–8.

Stringer, C. (2006) *Homo britannicus*. Penguin, London.

Tarasov, L. and Peltier, W. R. (2005) Arctic freshwater forcing of the Younger Dryas cold reversal. *Nature*, **435**, 662–5.

Tarasov, L. and Peltier, W. R. (2006) A calibrated deglacial drainage chronology for the North American continent: evidence of an Arctic trigger for the Younger Dryas. *Quaternary Science Reviews*, **25**, 659–88.

Teller, J. T. *et al.* (2005) Alternative routing of Lake Agassiz overflow during the Younger Dryas: new dates, paleotopography, and a re-evaluation. *Quaternary Science Reviews*, **24**, 1890–905.

Turney, C. S. M. *et al.* (2006) Climatic variability in the southwest Pacific during the Last Termination (20–10 ka BP). *Quaternary Science Reviews*, **25**, 886–903.

Turney, C. S. M. *et al.* (2007) Redating the advance of the New Zealand Franz Josef Glacier during the Last Termination: evidence for asynchronous climate change. *Quaternary Science Reviews*, doi: 10.1016/j.quascirev.2007.09.014.

Von Post, L. (1916) Einige südschwedischen Quellmoore. *Bulletin of the Geological Institute of Upsala*, **15**, 218–78.

Walter, K. M. *et al.* (2007) Thermokarst lakes as a source of atmospheric CH_4 during the last deglaciation. *Science*, **318**, 633–6.

Waters, M. R. and Stafford Jr, T. W. (2007) Redefining the age of Clovis: Implications for the peopling of the Americas. *Science*, **315**, 1122–6.

Chapter 7: Meltdown

Alley, R. B. (2000) *The Two-Mile Time Machine*. Princeton University Press, Princeton.

Alley, R. B. *et al.* (1997) Holocene climatic instability: a prominent, widespread event 8200 yr BP. *Geology*, **25**, 483–6.

Alley, R. B. *et al.* (1993) Abrupt increase in Greenland snow accumulation at the end of the Younger Dryas. *Nature*, **362**, 527–9.

Balter, M. (2007) Seeking agriculture's ancient roots. *Science*, **316**, 1830–5.

Barber, D. C. *et al.* (1999) Forcing of the cold event of 8,200 years ago by catastrophic drainage of Laurentide lakes. *Nature*, **400**, 344–8.

Beget, J. E. and Addison, J. A. (2007) Methane gas release from the Storegga submarine landslide linked to early Holocene climate change: a speculative hypothesis. *The Holocene*, **17**, 291–5.

Biello, D. (2006) Fact or fiction?: Archimedes coined the term 'Eureka!' in the bath. *Scientific American*, 8 December.

Blockley, S. M. (2005) Two hiatuses in human bone radiocarbon dates in Britain (17000 to 5000 cal BP). *Antiquity*, **79**, 505–13.

Blockley, S. P. E. *et al.* (2006) The chronology of abrupt climate change and Late Upper Palaeolithic human adaptation in Europe. *Journal of Quaternary Science*, **21**, 575–84.

Bondevik, S. (2003) Storegga tsunami sand in peat below the Tapes beach ridge at Harøy, western Norway, and its possible relation to an early Stone Age settlement. *Boreas*, **32**, 476–83.

Bryn, P. *et al.* (2005) Explaining the Storegga Slide. *Marine and Petroleum Geology*, **22**, 11–19.

Clarke, G. *et al.* (2003) Superlakes, megafloods, and abrupt climate change. *Science*, **301**, 922–3.

Crutzen, P. J. (2002) Geology of mankind. *Nature*, **415**, 23.

Crutzen, P. I. and Stoermer, E.F. (2000) The 'Anthropocene'. *IGBP Newsletter*, **41**, 12.

Davis, B. A. S. *et al.* (2003) The temperature of Europe during the Holocene reconstructed from pollen data. *Quaternary Science Reviews*, **22**, 1701–16.

Dawson, A. S. *et al.* (1990) Evidence for a tsunami from a Mesolithic site in Inverness, Scotland. *Journal of Archaeological Science*, **17**, 509–12.

Dyke, A. S. and Savelle, J. M. (2001) Holocene history of the Bering Sea bowhead whale (*Balaena mysticetus*) in its Beaufort Sea summer grounds off Southwestern Victoria Island, Western Canadian Arctic. *Quaternary Research*, **55**, 371–9.

Fisher, D. *et al.* (2006) Natural variability of Arctic sea ice over the Holocene. *Eos*, **87**, 273–5.

Fitch, S. *et al.* (2005) Late Pleistocene and Holocene depositional systems and the palaeogeography of the Dogger Bank, North Sea. *Quaternary Research*, **64**, 185–96.

Gupta, S. *et al.* (2007) Catastrophic flooding origin of shelf valley systems in the English Channel. *Nature*, **448**, 342–5.

Haflidason, H. *et al.* (2005) The dating and morphometry of the Storegga Slide. *Marine and Petroleum Geology*, **22**, 123–36.

Holland, M. *et al.* (2006) Future abrupt reductions in the summer Arctic sea ice. *Geophysical Research Letters*, **33**, doi:10.1029/2006GL028024.

Kaufman, D. S. *et al.* (2004) Holocene thermal maximum in the western Arctic (0–180° W). *Quaternary Science Reviews*, **23**, 529–60.

Klitgaard-Kristensen, D. *et al.* (1998) A regional 8200 cal. yr BP cooling event in northwest Europe, induced by final stages of the Laurentide ice-sheet deglaciation? *Journal of Quaternary Science*, **13**, 165–9.

Lambeck, K. (1995) Late Devensian and Holocene shorelines of the British Isles and North Sea from models of glacio-hydro-isostatic rebound. *Journal of the Geological Society of London*, **152**, 437–48.

Lambeck, K. and Chappell, J. (2001) Sea level change through the last glacial cycle. *Science*, **292**, 679–86.

Lavery, S. and Donovan, B. (2005) Flood risk management in the Thames Estuary looking ahead 100 years. *Philosophical Transactions of the Royal Society of London*, **363A**, 1455–74.

LeGrande, A. N. *et al.* (2006) Consistent simulations of multiple proxy responses to an abrupt climate change event. *Proceedings of the National Academy of Sciences*, **103**, 837–42.

Nadel, D. *et al.* (2004) Stone Age hut in Israel yields world's oldest evidence of bedding. *Proceedings of the National Academy of Sciences*, **101**, 6821–6.

Parfitt, S. A. *et al.* (2005) The earliest record of human activity in northern Europe. *Nature*, **438**, 1008–12.

Renfrew, C. (2006) Inception of agriculture and rearing in the Middle East. *Comptes Rendus Palevol.*, **5**, 395–404.

Richter-Menge, J. *et al.* (2006) *State of the Arctic Report*. NOAA OAR Special Report, NOAA/OAR/PMEL, Seattle, WA.

Rohling, E. J. and Pälike, H. (2005) Centennial-scale climate cooling with a sudden cold event around 8,200 years ago. *Nature*, **434**, 975–9.

Ruddiman, W. F. (2003) Orbital insolation, ice volume, and greenhouse gases. *Quaternary Science Reviews*, **22**, 1597–629.

Ruddiman, W. F. (2003) The Anthropogenic Greenhouse Era began thousands of years ago. *Climatic Change*, **61**, 261–93.

Ruddiman, W. F. (2005) *Plows, Plagues and Petroleum: How Humans Took Control of Climate*. Princeton University Press, Princeton.

Ruddiman, W. F. and Thomson, J. S. (2001) The case for human causes of increased atmospheric CH_4 over the last 5000 years. *Quaternary Science Reviews*, **20**, 1769–77.

Ruddiman, W. F. *et al.* (2005) A test of the overdue-glaciation hypothesis. *Quaternary Science Reviews*, **24**, 1–10.

Ruddiman, W. F. (2006) On 'The Holocene CO_2 rise: Anthropogenic or natural?'. *Eos*, **87**, 352–3.

Ryan, W. B. F. *et al.* (2003) Catastrophic flooding of the Black Sea. *Annual Review of Earth and Planetary Science*, **31**, 525–54.

Ryan, W. B. F. *et al.* (1997) An abrupt drowning of the Black Sea shelf. *Marine Geology*, **138**, 119–26.

Severinghaus, J. P. *et al.* (1998) Timing of abrupt climate change at the end of the Younger Dryas interval from thermally fractionated gases in polar ice. *Nature*, **391**, 141–6.

Sheldrick, C. *et al.* (1997) Palaeolithic barbed point from Gransmoor, East Yorkshire, England. *Proceedings of the Prehistoric Society*, **63**, 359–70.

Shennan, I. and Horton, B. (2002) Holocene land- and sea-level changes in Great Britain. *Journal of Quaternary Science*, **17**, 511–26.

Shennan, I. *et al.* (2006) Relative sea-level changes, glacial isostatic modelling and ice-sheet reconstructions from the British Isles since the Last Glacial Maximum. *Journal of Quaternary Science*, **21**, 585–99.

Shennan, I. *et al.* (2000) Modelling western North Sea paleogeographies and tidal changes during the Holocene. In *Holocene Land-Ocean Interaction and Environmental Change Around the North Sea*. Geological Society, London, Special Publications, 166, 299–319.

Siddall, M. *et al.* (2004) Testing the physical oceanographic implications of the suggested sudden Black Sea infill 8400 years ago. *Paleoceanography*, **19**, doi: 10.1029/2003PA000903.

Singh, S. (1999) Serendipity on the rebound. *The Independent*, 21 November.

Smith, D. E. *et al.* (2006) Towards improved empirical isobase models of Holocene land uplift for mainland Scotland, UK. *Philosophical Transactions of the Royal Society of London*, **364A**, 949–72.

Smith, D. E. *et al.* (2004) The Holocene Storegga Slide tsunami in the United Kingdom. *Quaternary Science Reviews*, **23**, 2291–321.

Stroeve, J. *et al.* (2007) Arctic sea ice decline: Faster than forecast. *Geophysical Research Letters*, **34**, doi:10.1029/2007GL029703.

Törnqvist, T.E. *et al.* (2004) Tracking the sea-level signature of the 8.2 ka cooling event: New constraints from the Mississippi Delta. *Geophysical Research Letters*, **31**, doi: 10.129/2004GL021429.

Turney, C. S. M. and Brown, H. (2007) Catastrophic early Holocene sea level rise, human migration and the Neolithic transition in Europe. *Quaternary Science Reviews*, **26**, 2036–41.

von Grafenstein, U. *et al.* (1999) A mid-European decadal-climate record from 15,500 to 5,000 years B.P. *Science*, **284**, 1654–7.

Wagner, B. *et al.* (2007) First indication of Storegga tsunami deposits from East Greenland. *Journal of Quaternary Science*, **22**, 321–5.

Waller, M. P. and Long, A. J. (2003) Holocene coastal evolution and sea-level change on the southern coast of England: a review. *Journal of Quaternary Science*, **18**, 351–9.

Chapter 8: Rise and fall

Ainsworth, E. A. and Long, S. P. (2005) What have we learned from fifteen years of Free Air Carbon Dioxide Enrichment (FACE)? A meta-analytic review of the responses of photosynthesis, canopy properties and plant production to rising CO_2. *New Phytologist*, **165**, 351–72.

Bonani, G. *et al.* (1994) AMS ^{14}C age determinations of tissue, bone and grass samples from the Ötztal Ice Man. *Radiocarbon*, **36**, 247–50.

Bond, G. *et al.* (1997) A pervasive millennial-scale cycle in North Atlantic Holocene and glacial climates. *Science*, **278**, 1257–66.

Bond, G. *et al.* (2001) Persistent solar influence on North Atlantic climate during the Holocene. *Science*, **294**, 2130–6.

Bowen M. (2005) *Thin Ice: Unlocking the Secrets of Climate in the World's Highest Mountains*. Henry Holt, New York.

British Broadcasting Corporation (1998) They called the wind Al Nino. http://news.bbc.co.uk/1/hi/world/americas/61579.stm.

Brooks, N. (2006) Cultural responses to aridity in the Middle Holocene and increased social complexity. *Quaternary International*, **151**, 29–49.

Claussen, M. *et al.* (1999) Simulation of an abrupt change in Saharan vegetation in the mid-Holocene. *Geophysical Research Letters*, **26**, 2037–40.

Coombes, P. and Barber, K. (2005) Environmental determinism in Holocene research: causality or coincidence? *Area*, **37**, 303–11.

Crabtree, G. W. and Lewis, N. S. (2007) Solar energy conversion. *Physics Today*, March, 37–42.

Darwin, C. (1846) An account of the fine dust which often falls on vessels in the Atlantic Ocean. *Quarterly Journal of the Geological Society*, **2**, 26–30.

deMenocal, P. *et al.* (2000) Abrupt onset and termination of the African Humid Period: rapid climate responses to gradual insolation forcing. *Quaternary Science Reviews*, **19**, 347–61.

Denton, G. H. and Karlén, W. (1973) Holocene climatic variations – their pattern and possible cause. *Quaternary Research*, **3**, 155–205.

Dyurgerov, M. B. and Meier, M. F. (2000) Twentieth century climate change: evidence from small glaciers. *Proceedings of the National Academy of Sciences*, **97**, 1406–11.

Foukal, P. *et al.* (2006) Variations in solar luminosity and their effect on the Earth's climate. *Nature*, **443**, 161–6.

Frank, N., Mangini, A. and Korfmann, M. (2002) ^{230}Th/U dating of the Trojan 'Water Quarries'. *Archaeometry*, **44**, 305–14.

Goelzer, H. *et al.* (2006) Tropical versus high latitude freshwater influence on the Atlantic circulation. *Climate Dynamics*, **27**, 715–25.

Grosjean, M. *et al.* (2007) Ice-borne prehistoric finds in the Swiss Alps reflect Holocene glacier fluctuations. *Journal of Quaternary Science*, **22**, 203–7.

Grove, R. H. (1998) Global impact of the 1789–93 El Niño. *Nature*, **393**, 318–19.

Hoerling, M. and Kumar, A. (2003) The perfect ocean for drought. *Science*, **299**, 691–4.

Keenan, J. (2005) Looting the Sahara: the material, intellectual and social implications of the destruction of cultural heritage (briefing). *The Journal of North African Studies*, **10**, 471–89.

Kuper, R. and Kröpelin, S. (2006) Climate-controlled Holocene occupation in the Sahara: motor of Africa's evolution. *Science*, **313**, 803–7.

Kutschera, W. and Rom, W. (2000) Ötzi, the prehistoric Iceman. *Nuclear Instruments and Methods in Physics Research*, **B164**, 12–22.

Kutzbach, J. E. (1981) Monsoon climate of the early Holocene: climatic experiment using the Earth's orbital parameters for 9000 years ago. *Science*, **214**, 59–61.

Kutzbach, J. E. and Liu, Z. (1997) Response of the African monsoon to orbital forcing and ocean feedbacks in the middle Holocene. *Science*, **278**, 440–4.

Mangini, A. (2007) Der Einfluss des Klimawandels auf die Siedlungsperioden von Troia. *The Studia Troica*, **17**, 1–6.

220 ICE, MUD AND BLOOD

McGregor, H. V. and Gagan, M. K. (2004) Western Pacific coral [18]O records of anomalous Holocene variability in the El Niño-Southern Oscillation. *Geophysical Research Letters*, **31**, doi: 10.1029/2004GL019972.

McPhaden, M. J. *et al.* (2006) ENSO as an integrating concept in earth science. *Science*, **314**, 1740–4.

Molloy, K. and O'Connell, M. (1995) Palaeoecological investigations towards the reconstruction of environment and land-use changes during prehistory at Céide Fields, western Ireland. *Probleme der Küstenforschung im südlichen Nordseegebiet*, **23**, 187–225.

Moy, C.M. *et al.* (2002) Variability of El Niño/Southern Oscillation activity at millennial timescales during the Holocene epoch. *Nature*, **420**, 162–5.

Muscheler, R. *et al.* (2007) Solar activity during the last 1000 yr inferred from radionuclide records. *Quaternary Science Reviews*, **26**, 82–97.

Nof, D. *et al.* (2006) Is there a paleolimnological explanation for 'walking on water' in the Sea of Galilee? *Journal of Paleolimnology*, **35**, 417–39.

O'Brien, S. R. *et al.* (1995) Complexity of Holocene climate as reconstructed from a Greenland ice core. *Science*, **270**, 1962–4.

Pernter, P. *et al.* (in press) Radiologic proof for the Iceman's cause of death (ca. 5,300 BP). *Journal of Archaeological Science*.

Porter, S. C. (2000) Onset of Neoglaciation in the Southern Hemisphere. *Journal of Quaternary Science*, **15**, 395–408.

Porter, S. C. and Denton, G. H. (1967) Chronology of Neoglaciation in the North America Cordillera. *American Journal of Science*, **265**, 177–210.

Prell, W. L. and Kutzbach, J. E. (1987) Monsoon variability over the past 150,000 years. *Journal of Geophysical Research*, **92**, 8411–25.

Rodbell, D. T. *et al.* (1999) An ~15,000-year record of El Niño-driven alluviation in southwestern Ecuador. *Science*, **283**, 516–20.

Ruddiman, W. F. *et al.* (2005) A test of the overdue-glaciation hypothesis. *Quaternary Science Reviews*, **24**, 1–10.

Sandweiss, D. H. *et al.* (2004) Geoarchaeological evidence for multidecadal natural climatic variability and ancient Peruvian fisheries. *Quaternary Research*, **61**, 330–4.

Sandweiss, D. H. *et al.* (1996) Geoarchaeological evidence from Peru for a 5000 years B.P. onset of El Niño. *Science*, **273**, 1531–3.

Thompson, L. G. *et al.* (1998) A 25,000-year tropical climate history from Bolivian ice cores. *Science*, **282**, 1858–64.

Thompson, L. G. *et al.* (2000) Ice-core palaeoclimate records in tropical South America since the Last Glacial Maximum. *Journal of Quaternary Science*, **15**, 377–94.

Thompson, L. G. *et al.* (1985) A 1500-year record of tropical precipitation in ice cores from the Quelccaya ice cap, Peru. *Science*, **229**, 971–3.

Thompson, L. G. *et al.* (2006) Abrupt tropical climate change: past and present. *Proceedings of the National Academy of Sciences*, **103**, 10536–43.

Thompson, L. G. *et al.* (2002) Kilimanjaro ice core records: evidence of Holocene climate change in tropical Africa. *Science*, **298**, 589–93.

Thompson, L. G. *et al.* (1995) Late glacial stage and Holocene tropical ice core records from Huascarán, Peru. *Science*, **269**, 46–50.

Tudhope, A. W. *et al.* (2001) Variability in the El Niño-Southern Oscillation through a glacial–interglacial cycle. *Science*, **291**, 1511–17.

Turney, C. *et al.* (2005) Testing solar forcing of pervasive Holocene climate cycles. *Journal of Quaternary Science*, **20**, 511–18.

Turney, C. S. M. and Hobbs, D. (2006) ENSO influence on Holocene Aboriginal populations in Queensland, Australia. *Journal of Archaeological Science*, **33**, 1744–8.

Turney, C. S. M. *et al.* (2006) Holocene climatic change and past Irish societal response. *Journal of Archaeological Science*, **33**, 34–8.

Vellinga, M. and Wu, P. (2004) Low-latitude freshwater influence on centennial variability of the Atlantic thermohaline circulation. *Journal of Climate*, **17**, 4498–511.

Walker, G. T. (1924) Correlation in seasonal variations of weather. IX. A further study of world weather. *Memoirs of the Indian Meteorological Department*, **24**, 275–332.

Walker, G. T. (1923) Correlation in seasonal variations of weather. VIII. A preliminary study of world-weather. *Memoirs of the Indian Meteorological Department*, **24**, 75–131.

Walker, G. T. (1936) Seasonal weather and its prediction. *Smithsonian Institution Annual Report for 1935*, 117–38.

Zagorski, N. (2006) Profile of Lonnie G. Thompson. *Proceedings of the National Academy of Sciences*, **103**, 11437–9.

Chapter 9: Droughts, vines and frost fairs

Arneborg, J. *et al.* (1999) Change of diet of the Greenland Vikings determined from stable carbon isotope analysis and ^{14}C dating of their bones. *Radiocarbon*, **41**, 157–68.

Babbage, C. (1838) *The Ninth Bridgewater Treatise, a Fragment*, 2nd edn. John Murray, London.

222 ICE, MUD AND BLOOD

Baroni, C. and Orombelli, G. (1996) The Alpine 'Iceman' and Holocene climatic change. *Quaternary Research*, **46**, 78–83.

Benson, L. *et al.* (2007) Anasazi (pre-Columbian Native American) migrations during the middle-12th and late-13th centuries – were they drought induced? *Climatic Change*, **83**, 187–213.

Brádzil, R. *et al.* (2005) Historical climatology in Europe – the state of the art. *Climatic Change*, **70**, 363–430.

Brooks, C. E. P. (1922) *The Evolution of Climate*. Benn Brothers, London.

Buckland, P. C. *et al.* (1996) Bioarchaeological and climatological evidence for the fate of Norse farmers in medieval Greenland. *Antiquity*, **70**, 88–96.

Chuine, I. *et al.* (2004) Grape ripening as a past climate indicator. *Nature*, **432**, 289–90.

Cook, E. R. *et al.* (2002) Evidence for a 'Medieval Warm Period' in a 1,100 year tree-ring reconstruction of past austral summer temperatures in New Zealand. *Geophysical Research Letters*, **29**, 10.129/2001GL014580.

Cook, E. R. *et al.* (2004) Long-term aridity changes in the western United States. *Science*, **306**, 1015–18.

Cook, E. R. *et al.* (2007) North American drought: Reconstructions, causes, and consequences. *Earth Science Reviews*, **81**, 93–134.

Crutzen, P. J. (2006) Albedo enhancement by stratospheric sulfur injection: A contribution to resolve a policy dilemma? *Climatic Change*, **77**, 211–20.

Curtis, J. H. *et al.* (1996) Climate variability on the Yucatan Peninsula (Mexico) during the past 3500 years, and implications for Maya cultural evolution. *Quaternary Research*, **46**, 37–47.

Dahl-Jensen, D. *et al.* (1999) Monte Carlo inverse modelling of the Law Dome (Antarctica) temperature profile. *Annals of Glaciology*, **29**, 145–50.

Dansgaard, W. *et al.* (1975) Climatic changes, Norsemen and modern man. *Nature*, **255**, 24–8.

deMenocal, P. B. (2001) Cultural responses to climate change during the late Holocene. *Science*, **292**, 667–73.

Diamond, J. (2005) *Collapse: How Societies Choose to Fail or Survive*. Penguin, London.

Douglass, A. E. (1919) *Climatic Cycles and Tree Growth I*. Carnegie Institute, Washington.

Douglass, A. E. (1929) The secret of the Southwest solved by talkative tree rings. *National Geographic*, **56**, 736–70.

Folan, W. J. *et al.* (1995) Calakmul: New data from an ancient Maya capital in Campeche, Mexico. *Latin American Antiquity*, 6, 310–34.

Forbes, J. D. (1832) Report upon the recent progress and present state of meteorology. *Report of the 2nd Meeting of the British Association for the Advancement of Science*, 2, 196–258.

Friedrich, M. *et al.* (2004) The 12,460-year Hohenheim oak and pine tree-ring chronology from central Europe – a unique annual record for radiocarbon calibration and paleoenvironment reconstructions. *Radiocarbon*, 46, 1111–22.

Gill, R. B. (2000) *The Great Maya Droughts: Water, Life and Death.* University of New Mexico Press, Albuquerque.

Goosse, H. *et al.* (2006) The origin of the European 'Medieval Warm Period'. *Climate of the Past*, 2, 99–113.

Haug, G. H. *et al.* (2001) Southward migration of the Intertropical Convergence Zone. *Science*, 293, 1304–8.

Hendy, E. J. *et al.* (2002) Abrupt decrease in tropical Pacific sea surface salinity at the end of Little Ice Age. *Science*, 295, 1511–14.

Hodell, D. A. *et al.* (1995) Possible role of climate in the collapse of classic Maya civilization. *Nature*, 375, 391–4.

Jones, P. D. and Mann, M. E. (2004) Climate over past millennia. *Reviews of Geophysics*, 42, doi: 2003RG000143.

Jones, P. D. *et al.* (2001) The evolution of climate over the last millennium. *Science*, 292, 662–7.

Lamb, H. H. (1965) The early Medieval warm epoch and its sequel. *Palaeogeography, Palaeoclimatology, Palaeoecology*, 1, 13–37.

Lund, D. C. *et al.* (2006) Gulf Stream density structure and transport during the past millennium. *Nature*, 444, 601–4.

Magnusson, M. (2003) *The Vikings.* Tempus, Stroud, UK.

Mann, M. E. (2007) Climate over the past two millennia. *Annual Review of Earth and Planetary Sciences*, 35, 111–36.

Mann, M. E. *et al.* (1995) Global interdecadal and century-scale climate oscillations during the past five centuries. *Nature*, 378, 266–70.

Mann, M. E. *et al.* (1998) Global-scale temperature patterns and climate forcing over the past six centuries. *Nature*, 392, 779–87.

Mann, M. E. *et al.* (1999) Northern Hemisphere temperatures during the past millennium: inferences, uncertainties, and limitations. *Geophysical Research Letters*, 26, 759–62.

Marlar, R. A. *et al.* (2000) Biochemical evidence of cannibalism at a prehistoric Puebloan site in southwestern Colorado. *Nature*, 407, 74–8.

Matthes, F. (1939) Report of Committee on Glaciers, April 1939. *Eos*, **20**, 518–23.

McKinzey, K. M. *et al.* (2004) A revised Little Ice Age chronology of the Franz Josef Glacier, Westland, New Zealand. *Journal of the Royal Society of New Zealand*, **34**, 381–94.

Meko, D. M. *et al.* (2007) Medieval drought in the upper Colorado River Basin. *Geophysical Research Letters*, **34**, doi:10.1029/2007GL029988.

Miller, J. F. (1849) On the meteorology of the Lake District of Cumberland and Westmoreland; including the results of experiments on the fall of rain at various heights above the Earth's surface, up to 3166 feet above the mean sea level. *Philosophical Transactions of the Royal Society of London*, **139**, 53–89.

Moberg, A. *et al.* (2005) Highly variable Northern Hemisphere temperatures reconstructed from low- and high-resolution proxy data. *Nature*, **433**, 613–17.

National Research Council (2006) *Surface Temperature Reconstructions for the Last 2,000 Years*. National Academy of Sciences, The National Academies Press, Washington.

Oppenheimer, C. (2003) Climatic, environmental and human consequences of the largest known historic eruption: Tambora volcano (Indonesia) 1815. *Progress in Physical Geography*, **27**, 230–59.

Panagiotakopulu, E. *et al.* (2007) Fossil insect evidence for the end of the Western Settlement in Norse Greenland. *Naturwissenschaften*, **94**, 300–6.

Polyak, V. J. and Asmerom, Y. (2001) Late Holocene climate and cultural changes in the southwestern United States. *Science*, **294**, 148–51.

Rampino, M. R. and Self, S. (1982) Historic eruptions of Tambora (1815), Krakatau (1883), and Agung (1963), their stratospheric aerosols, and climatic impact. *Quaternary Research*, **18**, 127–43.

Reiter, P. (2000) From Shakespeare to Defoe: Malaria in England in the Little Ice Age. *Emerging Infectious Diseases*, **6**, 1–11.

Reynolds, A. C. *et al.* (2005) $^{87}Sr/^{86}Sr$ sourcing of ponderosa pine used in Anasazi great house construction at Chaco Canyon, New Mexico. *Journal of Archaeological Science*, **32**, 1061–75.

Schmidt, G. A. *et al.* (2004) General circulation modelling of Holocene climate variability. *Quaternary Science Reviews*, **23**, 2167–81.

Seager, R. *et al.* (2007) Model projections of an imminent transition to a more arid climate in southwestern North America. *Science*, **316**, 1181–7.

Selley, R. C. (2004) *The Winelands of Britain: Past, Present and Prospective*. Petravin, Dorking, UK.

Stine, S. (1994) Extreme and persistent drought in California and Patagonia during mediæval time. *Nature*, **369**, 546–9.

Stothers, R. B. (1984) The great Tambora eruption in 1815 and its aftermath. *Science*, **224**, 1191–8.

Stott, P. A. *et al.* (2004) Human contribution to the European heatwave of 2003. *Nature*, **432**, 610–14.

Swade, D. (2001) *The Difference Engine: Charles Babbage and the Quest to Build the First Computer*. Viking, New York.

The Times (1834) Change of climate. *The Times*, 18 February, p. 3.

Trenberth, K. E. and Dai, A. (2007) Effects of Mount Pinatubo volcanic eruption on the hydrological cycle as an analog of geoengineering. *Geophysical Research Letters*, **34**, doi: 10.1029/2007GL030524.

Unwin, T. (1990) Saxon and early Norman viticulture in England. *Journal of Wine Research*, **1**, 61–75.

Wigley, T. M. L. (2006) A combined mitigation/geoengineering approach to climate stabilization. *Science*, **314**, 452–4.

Woodhouse, C. A. *et al.* (2006) Updated streamflow reconstructions for the Upper Colorado River Basin. *Water Resources Research*, **42**, doi:10.1029/2005WR004455.

Zorita, E. *et al.* (2004) Climate evolution in the last five centuries simulated by an atmosphere–ocean model: global temperatures, the North Atlantic Oscillation and the Late Maunder Minimum. *Meteorologische Zeitschrift*, **13**, 271–89.

Chapter 10: The rising tide

Boko, M. *et al.* (2007) Africa. *Climate Change 2007: Impacts, Adaptation and Vulnerability. Contribution of Working Group II to the Fourth Assessment Report of the Intergovernmental Panel on Climate Change* (eds. M. L. Parry *et al.*), pp. 433–67. Cambridge University Press, Cambridge UK.

Domack, E. *et al.* (2005) Stability of the Larsen B ice shelf on the Antarctic Peninsula during the Holocene epoch. *Nature*, **436**, 681–5.

Narisma, G. T. *et al.* (2007) Abrupt changes in rainfall during the twentieth century. *Geophysical Research Letters*, **34**, doi:10.1029/2006GL028628.

Rahmstorf, S. *et al.* (2007) Recent climate observations compared to projections. *Science*, **316**, 709.

Rignot, E. *et al.* (2004) Accelerated ice discharge from the Antarctic
    Peninsula following the collapse of Larsen B ice shelf. *Geophysical
    Research Letters*, **31**, doi:10.1029/2004GL020697.
Vaughan, D. G. (in press) West Antarctic Ice Sheet collapse-the fall and
    rise of a paradigm. *Climatic Change*.
Webster, J. M. and Davies, P. J. (2003) Coral variation in two deep drill
    cores: significance for the Pleistocene development of the Great
    Barrier Reef. *Sedimentary Geology*, **159**, 61–80.

INDEX

232   INDEX